Paleomicrobiology of Humans

Paleomicrobiology of Humans

Editors:

Michel Drancourt

Didier Raoult

ASM
PRESS

Washington, DC

Library of Congress Cataloging-in-Publication Data

Names: Drancourt, Michel (Professor), editor. | Raoult, Didier, editor.
Title: Paleomicrobiology of humans / editors, Michel Drancourt, Didier Raoult.
Description: Fifth edition. | Washington, DC : ASM Press, [2017] | Includes
  bibliographical references.
Identifiers: LCCN 2016035459 | ISBN 9781555819163 (paperback : alk. paper)
Subjects: LCSH: Paleomicrobiology. | Micropaleontology. | Paleopathology. |
  Communicable diseases--Epidemiology--History.
Classification: LCC RA643 .P245 2016 | DDC 614.4/2--dc23 LC record available at https://lccn.loc.
gov/2016035459

ISBN 978-1-55581-916-3
e-ISBN 978-1-55581-917-0
doi:10.1128/9781555819170

Printed in the United States of America

10 9 8 7 6 5 4 3 2 1

Address editorial correspondence to: ASM Press, 1752 N St., NW, Washington, DC 20036-2904, USA.
Send orders to: ASM Press, P.O. Box 605, Herndon, VA 20172, USA.
Phone: 800-546-2416; 703-661-1593. Fax: 703-661-1501.
E-mail: books@asmusa.org
Online: http://estore.asm.org

Contents

Contributors

Laurent Abi-Rached
Aix Marseille Université, URMITE, UMR CNRS 7278, IRD 198,
INSERM 1095, Faculté de Médecine, Institut Hospitalo-Universitaire
Méditerranée-Infection, Marseille, France

Gérard Aboudharam
Aix Marseille Université, URMITE, UMR CNRS 7278, IRD 198,
Inserm 1095, Faculté de Médecine, Marseille, France

Emmanouil Angelakis
Unité de Recherche sur les Maladies Infectieuses Transmissibles
et Emergentes, Aix-Marseille Université, UM63, CNRS 7278,
IRD 198, INSERM U1095, Marseille, France

Sandra Appelt
Aix Marseille Université, URMITE, UM63, CNRS 7278, IRD 198,
Inserm 1095, Marseille, France

Adauto Araujo
Escola Nacional de Saude Publica Sergio Arouca, Fundacao Oswaldo
Cruz, Brazil

Yassina Bechah
Unité de Recherche sur les Maladies Infectieuses Transmissibles
et Emergentes, Aix-Marseille Université, UM63, CNRS 7278,
IRD 198, INSERM U1095, Marseille, France

Philippe Biagini
Viral Emergence and Co-Evolution Unit, UMR 7268 ADES, Aix-Marseille
University / French Blood Agency / CNRS, Marseille, France

Saverio Caini
Institute for Cancer Research and Prevention, Unit of Molecular and
Nutritional Epidemiology, Florence, Italy

Dominique Castex
UMR 5199 du CNRS, PACEA, Anthropologie des Populations Passées
et Présentes, Pessac, France

Eric Crubézy
AMIS Laboratory, UMR 5288, CNRS / University of Toulouse /
University of Strasbourg, Toulouse, France

Helen Donoghue
Centre for Clinical Microbiology, Division of Infection and Immunity, University College London, United Kingdom

Rezak Drali
Unité de Recherche sur les Maladies Infectieuses et Tropicales Emergentes: URMITE, Aix Marseille Université, UMR CNRS 7278, IRD 198, INSERM 1095, Faculté de Médecine, Marseille, France

Michel Drancourt
Aix Marseille Université, URMITE, UM63, CNRS 7278, IRD 198, Inserm 1095, Marseille, France

Olivier Dutour
Laboratoire d'Anthropologie biologique Paul Broca – École Pratique des Hautes Etudes, PSL Research University Paris, Paris, France

Pierre-Edouard Fournier
Unité de Recherche sur les Maladies Infectieuses et Tropicales Emergentes (URMITE), UM63, CNRS7278, IRD198, Inserm 1095, Aix-Marseille Université, Marseille, France

Eduardo Gotuzzo
Institute of Tropical Medicine, Peruvian University Cayetano Heredia, Lima, Peru

Sacha Kacki
UMR 5199 du CNRS, PACEA, Anthropologie des Populations Passées et Présentes, Pessac, France

Matthieu Le Bailly
Franche-Comté University, CNRS UMR 6249 Chrono-Environment, Besançon, France

Donatella Lippi
Department of Experimental and Clinical Medicine, University of Florence, Florence, Italy

Kosta Mumcuoglu
Parasitology Unit, Department of Microbiology and Molecular Genetics, The Kuvin Center for the Study of Infectious and Tropical Diseases, Hadassah Medical School, The Hebrew University, Jerusalem, Israel

Andreas Nerlich
Institute of Pathology, Academic Clinic Munich-Bogenhausen, Munich, Germany

Abiola Olumuyiwa Olaitain
Aix Marseille Université, Unité de Recherche sur les Maladies Infectieuses et Tropicales Emergentes (URMITE), UMR CNRS 7278, IRD 198, INSERM 1095, Faculté de Médecine, Marseille, France

Didier Raoult
Aix Marseille Université, URMITE, UMR CNRS 7278, IRD 198,
INSERM 1095, Faculté de Médecine, Institut Hospitalo-Universitaire
Méditerranée-Infection, Marseille, France

Jean-Marc Rolain
Aix Marseille Université, Unité de Recherche sur les Maladies
Infectieuses et Tropicales Emergentes (URMITE), UMR CNRS 7278,
IRD 198, INSERM 1095, Faculté de Médecine, Marseille, France

Mauro Rubini
Department of Archaeology Foggia University, Foggia, Italy

Michel Signoli
UMR 7268 ADES, Anthropologie bio-culturelle, Droit, Ethique,
Santé - AMU/CNRS/EFS, Marseille, France

Mark Spigelman
Centre for Clinical Microbiology, Division of Infection & Immunity,
University College London, London, United Kingdom

Catherine Thèves
AMIS Laboratory, UMR 5288, CNRS / University of Toulouse /
University of Strasbourg, Toulouse, France

Stéfan Tzortzis
UMR 7268 ADES, Anthropologie bio-culturelle, Droit, Ethique,
Santé - AMU/CNRS/EFS, Marseille, France

Introduction

Paleomicrobiology is a new field that aims to identify past epidemics at the crossroads of different specialties such as anthropology, medicine, molecular biology and microbiology. Paleomicrobiology is facing several types of problems that are discussed in this book. On the one hand the recognition of human remains associated with epidemic outbreaks, the graves associated with disasters, demographic structures revealing the presence of epidemic moment (Chapters 1 and 2). On the other hand, paleomicrobiology, the history of epidemics, helps to understand the evolution of the history of human beings since now we can find the genetic markers associated with humans, like the gene HLA.LILR inherited from archaic hominids (Neanderthal or Denisovan man) in some populations that presumably have survived due to their resistance to some epidemic pathogens. Paleomicrobiology also helps to track the human migrations (3,4). The materials which can be used to make the diagnosis in paleomicrobiology include soft tissue when it comes to mummies, the arthropods, especially lice, bones and teeth. Use of the dental pulp as a source of genetic material was first used in paleomicrobiology before being used in human genetics (5). The utilization of the dental pulp as a source of DNA research by PCR molecular techniques was initially the subject of a controversy about the authenticity of the results. This controversy, which lasted more than 10 years, is resolved now. The polemics about the initial results concerning plague led to a general reflection on the plague pandemics, which in its turn led to a conclusion that the plague pandemics were probably provoked by the outbreaks of lice. Paleomicrobiology presents the evidence of common epidemics provoked by Bartonella quintana (which is known to be transmitted by lice) and Yersinia pestis which have perfectly demonstrated its role in epidemics and which was confirmed by contemporary plague cases (6). The dispute over the results of the paleomicrobiology led to outlining the identification and interpretation criteria (7). Then, paleomicrobiology developed in different research areas, particularly in the analysis of human coprolites (8), the identification of antibiotic resistance in ancient samples that preceded by several million years the use of antibiotic (9), the history of epidemic typhus (10), of Bartonelose (12) tuberculosis (13), leprosy (14), former intestinal parasites (15), malaria (16), smallpox (17), cholera (18) and finally the history of human lice (19). As a final point, the anatomical analysis of ancient samples also plays a role in the identification of past disease. Altogether, this is the first complete comprehensive book updating (reporting on) the approach of a new multidisciplinary scientific field.

Acknowledgments

The Editors acknowledge their assistant Olga Cusack, IHU Mediterranée Infection, for the expert technical assistance in preparing this edition.

Demographic Patterns Distinctive of Epidemic Cemeteries in Archaeological Samples

1

DOMINIQUE CASTEX[1] and SACHA KACKI[1]

Some ancient burial grounds are valuable testimonies of past epidemics. For both funerary archaeologists and paleobiologists, such archaeological sites offer a remarkable research framework to reveal unfamiliar aspects of these historical events, especially those that occurred during periods for which very few or no written sources exist. The multiplicity of congresses, articles, and syntheses on this topic in recent years illustrates the interdisciplinary research that has progressively emerged over the past two decades.

The detection of an ancient mortality crisis generally relies on archaeological evidence such as mass graves—that is, pits where numerous bodies were buried at the same time (1). However, retaining the hypothesis of an epidemic crisis also requires some paleobiological investigations. First of all, a careful examination of the skeletal remains must be undertaken to make sure that they do not exhibit any traumatic lesions evocative of violent death. Such stigmata would be expected in the case of mass graves related to war or massacre, so that their absence favors the epidemic hypothesis. However, a paleopathogical study cannot demonstrate an epidemic context because the highly virulent pathogens involved kill individuals before any skeletal lesion can develop. Other methods of analysis are therefore needed to ascertain the precise nature of an epidemic. Besides paleomicrobiological studies, analysis of the age and

[1]UMR 5199 du CNRS, PACEA, Anthropologie des Populations Passées et Présentes, Pessac, France.
Paleomicrobiology of Humans
Edited by Michel Drancourt and Didier Raoult
© 2016 American Society for Microbiology, Washington, DC
doi:10.1128/microbiolspec.PoH-0015-2015

sex composition of skeletal assemblages can provide some valuable arguments. Indeed, the age- and sex-specific mortality rates within a population or part of a population at a given time and place may differ depending on the pathogen involved. The study of these biological parameters is particularly relevant to a discussion of the cause of an epidemic mortality crisis.

The scope of this article is voluntarily limited to the demographic analysis of burial grounds related to plague epidemics. Through the study of several of these sites, we illustrate the specificities of mortality by age and sex in such a context. We also discuss some demographic differences between several epidemic outbreaks.

MATERIALS AND METHODS

This work is based on a study of seven European burial sites of plague victims (Fig. 1). With the exception of the Hereford site, all of these burial grounds were used exclusively to bury the victims of plague epidemics. We therefore included in the analysis every skeleton recovered, whether from individual burials or multiple burials. For the Hereford cemetery, which was also used out of the epidemic context over a long period, we studied only the skeletons recovered from three multiple burials, which were proved to be the skeletons of the victims of a plague epidemic.

Two of the sites investigated (Sens and Poitiers) are contemporaneous with the first plague pandemic, known as Justinian's plague. This pandemic began in the 6th century and continued intermittently in Europe until the mid-8th century. Three other sites (Dreux, Hereford, and Barcelona) are contemporaneous with the Black Death—that is, the first outbreak of the second plague pandemic in the mid-14th century. Plague then waxed and waned in Europe until the late 18th century, and the last two sites we analyzed (Les Fédons and Dendermonde) are related to resurgences of the disease at the end of the 16th century. For every one of these archaeological burial

Sites	Chronology	References
1 - Sens (France)	6th c.	2
2 - Poitiers (France)	6th c.	3
3 - Hereford (England)	14th c.	4
4 - Dreux (France)	14th c.	2
5 - Barcelona (Spain)	14th c.	5
6 - Les Fédons (France)	16th c.	6
7 - Dendermonde (Belgium)	16th c.	7

○ Justinian's plague   ● Black Death   ◎ Resurgences

FIGURE 1 **Names and locations of the seven European burial sites used for the demographic analysis.**

grounds (Fig. 2), the cause of death is attested to by historical sources (Poitiers, Dendermonde) or by DNA analysis (Sens, Dreux, Hereford, Barcelona, Les Fédons).

The first step of the study of each site was to estimate the age at death and determine the sex of all the exhumed individuals by using reliable skeletal indicators (for a critical review of current methods and their application, see reference 8). Estimation of the age at death of non-adults was based primarily on the degree of dental maturation (9–11). When teeth were absent or fully mineralized, we additionally considered the diaphyseal length of the long bones (12) and the stage of fusion of secondary ossification centers (13, 14). Sex

determination was carried out only for adult individuals, with reliance primarily on the morphoscopic and morphometric characteristics of the hip bones (15, 16). In order to reduce the number of individuals whose sex could not be determined, we also performed, whenever possible, a secondary sex determination based on the extrapelvic dimorphism of each skeletal assemblage (17). These data were then used to define the composition of the population as well as possible. The mortality profile and the rate of masculinity (i.e., ratio of the number of men to the number of men and women) of each skeletal assemblage were critically discussed within the framework of a comparative analysis,

A

B

C

FIGURE 2 **Examples of burials with simultaneous deposits (i.e., mass graves): Poitiers, Poitou-Charentes, France (a); Dreux, Eure-et-Loir, France (b); Dendermonde, Belgium (c).**

with reliance on historical data and various standard life tables. Data were compared with an archaic/pre-Jennerian mortality pattern, with the variability of mortality and the confidence intervals of the probabilities of death taken into account. The model life tables of Ledermann (18) were used to make the comparisons. Such an approach has an objective distinct from those of traditional paleodemographic analyses, as it does not aim to determine the demographic characteristics of living populations (e.g., life expectancy at birth, average age at death) but rather to detect anomalous demographic patterns that can be interpreted as evidence of burial selection or abnormal mortality (2). If one assumes that the reference pattern(s) reflect a "natural" demography (i.e., deaths over a long period of time with no major mortality crises), the discrepancies between the reference and the paleodemographic data can reflect change in mortality parameters due to a peculiar event.

Given the poor accuracy of the methods of estimating adult age from the skeleton (8), mortality profiles were analyzed only for immature subjects. As in the model life tables used for comparison, the immature individuals were distributed into 5-year groups (apart from the first two groups, which were 1- and 4-year groups) of attained age. The mortality profile of the subjects from each site was then established by calculating the mortality quotients (Table 1) and was compared with the theoretical mortality pattern (18). The mortality quotient takes the form aQx, where x is the age of entry into an age group and a is the duration in years of the age group. aQx is the ratio of the number of people in an age group who die to the number of people alive at the start of that age group. It thus represents the probability of dying within a precise period of life. It differs from the rate of death, which is the proportion of deaths within an average population. The mortality quotient is more pertinent for comparing a theoretical natural mortality with the mortality profile obtained from an archaeological sample. The life tables of Ledermann (18) allow calculation of the 95% confidence interval of the mortality quotients (illustrated by a range of values in the graphs). The model life tables used in the current study are for a life expectancy at birth ranging from 25 to 35 years; these parameters lie between the 20- and 40-year life expectancy range that characterizes known pre-Jennerian populations (19, 20).

Rates of masculinity were calculated only for adults, as there is currently no reliable method of sex determination for sub-adults. Values were compared with a theoretical rate of 50%.

DEMOGRAPHIC CHARACTERISTICS OF THE PLAGUE BURIAL SITES

The results described below point out the demographic characteristics of each site. Com-

TABLE 1 Demographic data (number of individuals, age, sex, mortality quotients, juvenility index) for each plague site

	Sens		Poitiers		Hereford		Dreux		Barcelona		Les Fédons		Dendermone	
	N	aQx	N	aQx	N	aQx	N	aQx	N	aQx	N	aQx	N	aQx
0	0	0.00	0	0.00	5	27.03	2	28.98	3	25.00	4	30.08	0	0.00
1–4	3	41.10	1	18.87	23	127.78	15	223.88	12	102.56	19	147.29	5	50.51
5–9	10	142.90	4	76.92	28	178.34	10	192.31	16	152.38	24	218.18	22	234.04
10-14	8	133.30	9	187.50	14	108.53	5	119.05	8	89.89	19	220.93	17	236.11
15–19	7	134.60	8	205.13	9	78.26	5	135.14	11	135.80	7	104.48	11	200.00
20 and +	45		31		106		32		70		60		44	
Total males	14		10		40		11		6		30		20	
Total females	17		9		52		15		7		26		18	
Total number	73		53		185		69		120		133		99	
D5-14 / D20+	0.40		0.42		0.40		0.47		0.37		0.72		0.89	

parisons among the sites reveal similarities in mortality profiles specific for plague mortality, as well as some inter-site differences.

Proportion of Immature Individuals

According to the model life tables of Ledermann (18), the proportion of immature individuals in an archaic population theoretically ranges from 45% to 64%, for a life expectancy at birth of between 25 and 35 years. The values in the skeletal assemblages show some variation (Fig. 3). For three of the sites, the proportions recorded are compatible with the one expected in a context of natural mortality. For the four other sites, the values are slightly above the theoretical range. There is a small deficit of subjects who died before 20 years of age.

At first sight, the proportion of non-adults appears more important over time, from the first wave of Justinian's plague in the 6th century to the resurgences of the second pandemic in the 16th century. However, two sites from the 14th century (Hereford and Barcelona) differ from this general pattern. The differences in the proportion of imma-

ture individuals for the same time period may reflect geographical variations in the epidemiological characteristics of plague, resulting in a lower mortality rate in the youngest in Barcelona (41.7%) or a higher rate in Dreux (50.7%). However, we cannot exclude a potential bias that would have led to an underestimation of the number of immature individuals in the English and Spanish sites.

Two important biases can be mentioned. The first one concerns the Hereford cemetery. For this site, only individuals from mass graves were taken into account. Several studies have demonstrated that various burial methods were adopted during epidemics and that single graves may coexist with mass graves (6, 21). Thus, a consideration of only mass graves may yield misleading results. The second bias concerns the plague pit of Barcelona. Several of the 120 recovered skeletons consisted only of lower limb bones and were categorized as probable adults according the bone dimensions. This arbitrary choice may have resulted in an underestimation of the number of adolescents and consequently of the number of immature individuals as a whole.

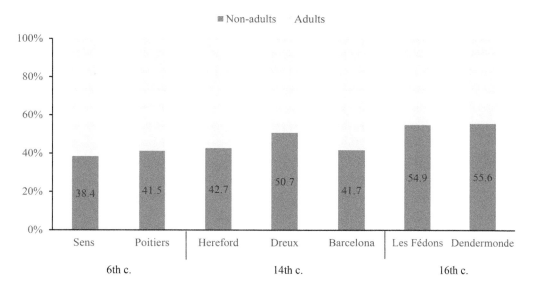

FIGURE 3 **Proportion of non-adults for each plague site.**

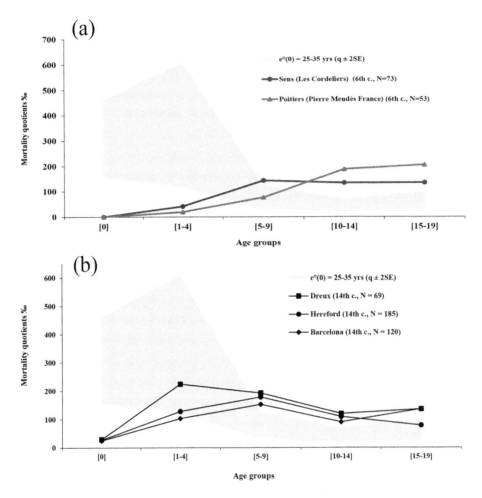

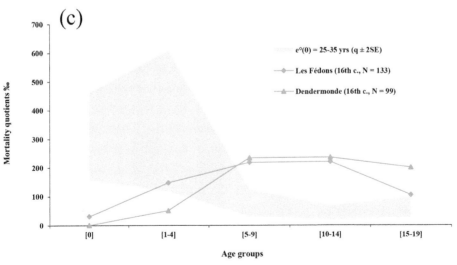

Based on the data currently available, the lowest representation of non-adults in the most recent plague burial sites deserves to be pointed out but cannot be interpreted as clear evidence of an evolution of plague epidemiological characteristics. Further studies would be needed to reach a conclusion on this question.

Mortality Quotients by Age Groups

The mortality profiles were established by calculating the mortality quotients for all age groups. They were then compared with a theoretical mortality model (18). Regardless of their chronology, the skeletal assemblages show recurrent anomalies (Fig. 4). In each case, there is a deficit in the number of children younger than 5 years of age. The infantile (younger than 1 year) mortality quotient and the mortality quotient of the 1- to 4-year age group are low in comparison with those characterizing a natural mortality. The most significant deficit affects the first age group. Whatever the site, the mortality quotient values of children younger than 1 year of age never exceed 30‰, whereas the theoretical values range from 156.3‰ to 459.8‰. For the two sites linked to Justinian's plague (Sens and Poitiers), infants are totally missing (Fig. 4a).

The relative constancy of the infantile mortality quotient values suggests that this deficit is specific to plague mortality. We cannot ignore that a cultural bias may have contributed to this anomaly, as very young children may have been buried in places other than the plague cemeteries. However, the anomaly is so recurrent that it is more likely that it is a distinguishing demographic characteristic of plague. This particularity should be investigated further in future research.

The mortality quotient of the 1- to 4-year age group is much more variable. Its value should range between 116.9‰ and 607.6‰. It is considerably lower for the two sites contemporaneous with Justinian's plague and for the Dendermonde cemetery, whereas it is close to the lower boundary of the theoretical interval for Barcelona, Hereford, and Les Fédons and has a median value for Dreux. Except for the last site, this mortality quotient never exceeds 147.3‰.

Conversely, the mortality quotients in the 5- to 9-year, 10- to 14-year, and 15- to 19-year age groups are noticeably higher than expected. Moreover, the proportions of individuals in all of these three age groups differ from the proportion for a natural mortality, which is characterized by a minimum number of deaths in the 10- to 14-year age group. These anomalies are obvious on the mortality profiles. The over-representation of older children and adolescents is also clearly demonstrated by the high values of the juvenility index (D_{5-14}/D_{20+}), which systematically exceed the range of 0.1 to 0.3 expected in an archaic mortality pattern (22). The juvenility indices are notably similar for plague burial sites from the 6th and 14th centuries, with values between 0.37 and 0.47 (Table 1). This anomaly is much more pronounced for the 16th century sites, which have very high juvenility indices.

Numerous individuals between the ages of 15 and 19 years also died. Although the mortality quotients for this age group vary depending on the sites, the values obtained are systematically close to the upper boundary of the theoretical interval (Hereford, Les Fédons) or far beyond it. This systematic surplus reveals that older adolescents were not spared during plague epidemics. No differences were observed according to the chronology of the epidemics.

FIGURE 4 **Distribution of ages at death for immature subjects. Comparisons with theoretical values of Ledermann (1969): Justinian's plague, 6th century (a); Black Death, 14th century (b); resurgences, 16th century (c).**

To summarize all these results, the age composition of the skeletal samples from plague burial sites clearly differs from the one expected in a non-epidemic context in two main respects: an under-representation of very young infants and a high number of deaths in individuals between the ages of 5 and 14 years. This conclusion confirms those of previous studies by showing a specific mortality profile that mimics the age structure of the living population (2). The current study, however, reveals how these demographic features vary from site to site and according to epidemic cycles. These new data open perspectives to be explored in future studies.

Adult Sex Distribution

At the level of the entire corpus, the sex proportions do not differ significantly from a theoretical masculinity rate of 50%. There are, however, some peculiarities according to the funerary site considered (Fig. 5). Two skeletal assemblages from the 14th century (Hereford and Dreux) are characterized by a larger proportion of females (masculinity rates of 43.5% and 42.3% respectively), whereas in others the sex ratios are more balanced. The disparity between the masculinity rate and its theoretical value is more important for the sites related to the Black Death than in those related to Justinian's plague and to the resurgences of the second pandemic; however, it must be stressed that the number of individuals whose sex could be determined is relatively small. Thus, there is no strong argument to conclude that there was a preferential female mortality during the 14th century plague epidemics. Given that the differences between the observed and theoretical masculinity rates are never significant, it seems that plague did not kill discriminately with respect to the sex of individuals.

DEMOGRAPHIC SPECIFICITIES OF THE PLAGUE

The study of seven burial sites related to plague revealed many similarities in the age

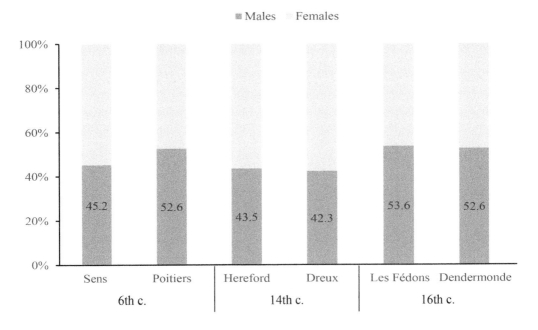

FIGURE 5 **Adult sex distribution for each plague site. Sex ratio of the Barcelona sample is unknown due to the low number of sexed individuals.**

and sex composition of the skeletal samples, which differ in some respects from those expected in a context of natural mortality. The results strongly reinforce those of previous anthropological studies (2, 23, 24, 25). The opportunity to analyze a large number of plague burial sites from different time periods has allowed a finer analysis of the demographic repercussions of the disease. This study reasonably supports that a demographic signature existed characterizing plague mortality, irrespective of chronological and geographical contexts. Its main characteristics are as follows:

1. mortality quotient for individuals younger than 20 years compatible with that for a natural mortality;
2. deficit of young children, particularly noticeable for individuals younger than 1 year;
3. excessive number of individuals between 5 and 14 years of age (high juvenility index);
4. over-representation of the oldest adolescents (15 to 19 years);
5. balanced sex ratio.

Despite these characteristics common to the sites, it is noteworthy that there are also some differences between sites. Minor variations may be ascribed to differences in the biological composition of the living populations. Others can be linked to methodological issues resulting, for example, from poor bone preservation or from incomplete recovery of skeletons due to partial excavation of the sites.

In contrast, some differences seem to be linked to the chronology of the funerary sites. The most significant one concerns the number of individuals in the 5- to 14-year age group. The extent to which they are over-represented varies considerably according to chronology. The anomaly is moderate during Justinian's plague and the Black Death, whereas it is much more pronounced during the resurgences of the second pandemic. The reason for these differences cannot be readily explained. There are currently few sites available for each period. Part of the explanation may perhaps be found in the biological composition of the human groups affected, in which individuals may already have been selected according to age. Another hypothesis is that the demographic impact of plague changed over time as the result of a modification in the virulence of *Yersinia pestis* and/or changes in the vulnerability of the human population to the disease.

CONCLUSIONS

By using adapted methodology and tools, paleobiologists can analyze the age and sex composition of an archaeological population and reveal its demographic particularities. Such an approach is extremely valuable in the interpretation of epidemic mortality crises, as it can provide arguments on which to base a discussion of the nature of the disease. As illustrated herein, it is possible to distinguish among different "anomalous demographic patterns" a specific plague mortality profile. The increasing number of burial sites now available makes it possible to explore new lines of research, such as tracking a possible evolution in the demographic signature of plague epidemics over centuries. The interpretation of such slight variations in the mortality profiles, however, will be possible only through the study of many more plague cemeteries all over Europe that are of various chronologies.

ACKNOWLEDGMENTS

This research has been funded by a grant from the Maison des Sciences de l'Homme d'Aquitaine (projet Région Aquitaine) and by a ministerial grant from the French Research National Agency (program of investments for the future, grant ANR-10-LABX-52). The anthropological study of the burials sites referred to in the paper was conducted as part of the research for a

Ph.D. thesis (S.K.) and as academic research (D.C.). S.K. is grateful to the institutions that provided access to the skeletal material from Hereford (Bradford Archaeological Research Centre, Bradford, United Kingdom), Barcelona (Museu d'Historia de Barcelona, Spain), Les Fédons (UMR 7268 – ADES, Marseille, France), and Dendermonde (Royal Belgian Institute of Natural Sciences, Brussels, Belgium).

CITATION

Castex D, Kacki S. 2016. Demographic patterns distinctive of epidemic cemeteries in archaeological samples. Microbiol Spectrum 4(4):PoH-0015-2015.

REFERENCES

1. **Castex D, Kacki S, Réveillas H, Souquet-Leroy I, Sachau-Carcel G, Blaizot F, Blanchard P, Duday H.** 2014. Revealing archaeological features linked to mortality increases. *Anthropologie (Brno)* **52**:299–318.
2. **Castex D.** 2008. Identification and interpretation of historical cemeteries linked to epidemics, p 23–48. *In* Raoult D, Drancourt M (ed), *Paleomicrobiology: Past Human Infections.* Springer-Verlag, Berlin, Germany.
3. **Godo C.** 2010. *Les inhumations intra-muros de Poitiers entre le IVe et le VIIe s. Biologie, gestes funéraires et essai d'interprétation.* Master thesis. University of Nanterre, Paris, France
4. **Stone R, Appleton-Fox N.** 1996. *A view from Hereford's past: a report on the archaeological excavation in Hereford Cathedral Close in 1993.* Logaston Press, Hereford, UK.
5. **Kacki S, Castex D.** 2014. La sépulture multiple de la basilique des Saints Martyrs Just et Pastor : bio-archéologie des restes humains. *Quaderns d'Arqueologia i Història de la Ciutat de Barcelona.* **10**:180–199.
6. **Bizot B, Castex D, Reynaud P, Signoli M.** 2005. *La saison d'une peste (avril-septembre 1590). Le cimetière des Fédons à Lambesc.* CNRS Éditions, Paris, France.
7. **Kacki S.** 2016. *Influence de l'état sanitaire des populations anciennes sur la mortalité en temps de peste : contribution à la paléoépidémiologie.* PhD thesis. University of Bordeaux, Bordeaux, France.
8. **Bruzek J, Schmitt A, Murail P.** 2005. Identification biologique individuelle en paléo-anthropologie. Détermination du sexe et estimation de l'âge au décès à partir du squelette, p 217–246. *In* Dutour O, Hublin JJ, Vandermeersch B (ed), *Origine et évolution humaine.* Comité des Travaux Historiques et Scientifiques, Paris, France.
9. **Moorrees CFA, Fanning EA, Hunt EE.** 1963a. Age variation of formation stages for ten permanent teeth. *J Dent Res* **42**:1490–1502.
10. **Moorrees CFA, Fanning EA, Hunt EE.** 1963b. Formation and resorption of three deciduous teeth in children. *Am J Phys Anthropol* **21**:205–213.
11. **Ubelaker DH.** 1978. *Human Skeletal Remains: Excavation, Analysis, Interpretation.* Aldine, Chicago, IL.
12. **Maresh MM.** 1970. Measurements from roentgenograms, p 157–200. *In* McCammon RW (ed), *Human Growth and Development.* Charles C. Thomas, Springfield, IL.
13. **Scheuer L, Black S.** 2000. *Developmental Juvenile Osteology.* Elsevier Academic Press, London, UK.
14. **Coqueugniot H, Weaver TD, Houët F.** 2010. Brief communication: a probabilistic approach to age estimation from infracranial sequences of maturation. *Am J Phys Anthropol* **142**:655–664.
15. **Bruzek J.** 2002. A method for visual determination of sex, using the human hip bone. *Am J Phys Anthropol* **117**:157–168.
16. **Murail P, Bruzek J, Houët F, Cunha E.** 2005. DSP: a tool for probabilistic sex diagnosis using worldwide variability in hip-bone measurements. *Bulletins et Mémoires de la Société d'Anthropologie de Paris* **17**:167–176.
17. **Murail P, Bruzek J, Braga J.** 1999. A new approach to sexual diagnosis in past populations. Practical adjustments from Van Vark's procedure. *Int J Osteoarchaeol* **9**:39–53.
18. **Ledermann S.** 1969. *Nouvelles tables-types de mortalité.* INED, PUF (Travaux et Documents, 53), Paris, France.
19. **Masset C.** 1975. La mortalité préhistorique, p 63–90. *In Cahiers du Centre de Recherches Préhistoriques, Université de Paris I, 4.* University of Paris, Paris, France.
20. **Sellier P.** 1996. La mise en évidence d'anomalies démographiques et leur interprétation: population, recrutement et pratiques funéraires du tumulus de Courtesoult, p 188–200. *In* Piningre JF (ed), *Le tumulus de Courtesoult (Haute-Saône) et le Ier Age du Fer dans le Bassin Supérieur de la Saône.* Éditions de la Maison des Sciences de l'Homme (DAF), Paris, France.
21. **Kacki S, Castex D, with collaboration of Cabezuelo U, Donat R, Duchesne S, Gaillard**

A. 2012. Réflexions sur la variété des modalités funéraires en temps d'épidémie. L'exemple de la Peste noire en contextes urbain et rural. *Archéologie Médiévale* **42:**1–21.

22. **Bocquet J-P, Masset C.** 1977. Estimateurs en paléodémographie. *L'Homme* **17:**65–90.

23. **Margerison BJ, Knüsel CJ.** 2002. Paleo-demographic comparison of a catastrophic and an attritional death assemblage. *Am J Phys Anthropol* **119:**134–148.

24. **Gowland RL, Chamberlain AT.** 2005. Detecting plague: palaeodemographic characterisation of a catastrophic death assemblage. *Antiquity* **79:**146–157.

25. **Signoli M.** 2006. *Études anthropologiques des crises démographiques en contexte épidémique: aspects paléo- et biodémographique de la peste en Provence.* Archaeopress (British Archaeological Reports, International Series, 1515), Oxford, UK.

Characterization of the Funeral Groups Associated with Plague Epidemics

2

STÉFAN TZORTZIS[1] and MICHEL SIGNOLI[1]

FUNERAL MANAGEMENT OF THE EXTRAORDINARY MORTALITY DURING THE PLAGUE

There are several scenarios regarding how burial sites in archaeological con-texts are discovered. We will focus on two scenarios according to the degree of historical knowledge regarding the studied sector. The excavation may be performed in a known funeral place or a highly suspected place (e.g., the interior or immediate exterior space in a religious monument or a parish cem-etery). Also, the excavation of unexpected graves or graves discovered by chance may occur in places that had unknown or forgotten funeral purposes.

In the first scenario, the discovery of funeral places associated with normal death management in its continuity (e.g., individual graves, family graves, and ossuaries) suggests that other funeral places should have signs that question the intensity of the mortality at the moment of their execution and the episodic nature of their use.

In the second scenario, the funeral structures may correspond to typical death management in an ancient chronological context. In contrast, these structures may correspond to a particular context of mortality, which would explain their location outside the areas routinely devoted to death manage-ment.

[1]UMR 7268 ADES, Anthropologie bio-culturelle, Droit, Ethique, Santé - AMU/CNRS/EFS, Marseille, France.
Paleomicrobiology of Humans
Edited by Michel Drancourt and Didier Raoult
© 2016 American Society for Microbiology, Washington, DC
doi:10.1128/microbiolspec.PoH-0011-2015

In both scenarios, the existence of burial places highlights the issue of identifying cases of abnormally high death rates that may have multiple origins (e.g., epidemics, natural disasters, or massive violence).

Given our 20-year experience of conducting excavations, we are confident when describing a number of funeral contexts that we associate with plague epidemics. The plague epidemic of 1720-1722 in Provence provided four examples of burials sites in two neighboring communities that represent the epidemic contexts of a peak and a relapse. There are two burial sites in Marseille: the massive grave of Observance (epidemic relapse in the spring of 1722) (1) and the grave of the "Major" cathedral (epidemic peak in September of 1720) (Fig. 1) (2). In Martigues, there are trenches located at the following sites: Capucins de Ferrières and the Délos

FIGURE 1 Grave in the "Major" cathedral in Marseille: the view southward (copyright INRAP).

(epidemic peak in the winter of 1720-1721) (3, 4).

There are similarities among the observed funeral practices for these four burial sites. These practices are evident within a minimalist general perspective as a result of the appearance of mass death management characteristics, such as multiple graves, a lack of standard positioning of the corpses, and the use of lime. However, the contextual data are not identical across all the sites. In Marseille, the excavated plague graves are situated inside the old town (i.e., *intra muros*). In the "Observance" site, the grave is composed of the corpses of the hospital staff (i.e., from the "Charité" and "Observance" hospitals), whereas the "Major" cathedral site consists of corpses from the entire city. In Martigues, the origin of the dead is more varied: in the "Capucins de Ferrières" site, corpses came from both the hospital and the city. This difference explains the absence of personal staff devoted to death management in the "Observance" site. The context for the "Observance" scenario is an epidemic relapse, which is supported by the following characteristics of the funeral procedures: the on-surface and simultaneously deep graves, which required coercive commitment from a large labor force to dig within a relatively limited period of time. Such graves could be utilized during consequent time periods, which may encompass the entire epidemic episode. The grave sites had to be sizeable in order to respond to the corpse removal needs of the care institute. As such, the trenches for both sites in Martigues, which are situated outside the city walls, have less depth than the graves in the "Observance" site, which were easier to prepare during that time period. At the "Delos" and "Capucins de Ferrières" sites, the graves had to be increased in number and most likely enlarged because of the growing need for burial space. This technical choice was most likely due to a pronounced concern on the part of the local authorities to act quickly without knowing the total number of corpses that the disease would generate. The burial site in "Observance" reveals concern dictated by the fast-paced events, including signs of urgent death management at the burial site in the "Major" church and execution time concerns at the graves in the "Capucin de Ferrières" and "Délos" sites. These concerns caused different yet relatively similar sepulchral configurations, given the simple characteristics of the funeral procedures.

In addition to the symbolic Provencal epidemics that were evident at the beginning of the 18th century, there are also sites in Fédons (epidemic of 1590, municipality of Lambesc) (5) and Lariey (epidemic of 1630, municipality of Puy-Saint-Pierre) (6). When faced with the plague epidemic, communities in Lambesc and Puy-Saint-Pierre organized hospices to take care of the patients. These hospital structures were associated with the burial spaces for the victims. At first glance, the cemetery for the hospice in Fédons did not differ from a parish cemetery, as there were numerous individual graves arranged in a rational way with corpses buried on their backs. However, the absence of a religious building (e.g., a church or chapel) near this cemetery and the presence of 21 double graves, in addition to four triple graves and one quadruple grave, indicate that this cemetery was indeed a crisis burial area (7). This hypothesis is supported by historical researches and molecular biology (8). The site in Lariey provides several originalities. This funeral site is situated outside the parish cemetery and the municipal territory. It is topographically identified as being from the 17th century (e.g., it is limited by stone walls) and was continually used over a long time period (according to Napoleon's land registry) until the 20th century, when it became a commemorative site (i.e., a milestone for annual pilgrimages). The spatial organization of the graves at these sites represents the plague progression, as follows: individual burials of the first victims, double burials due to the increase in mortality, and multiple burials associated with the peak of the disease (Fig. 2).

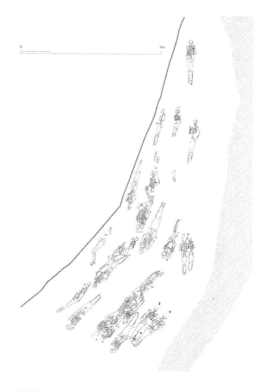

FIGURE 2 **General overview of the cemetery in Lariey—distribution of graves (drafted by B. B. Bizot and M. Olive; computer-aided design by B. B. Bizot).**

This progression has been confirmed through an analysis of historical archives and the biological detection of the F1 antigen specific for *Yersinia pestis* (9).

Finally, to highlight that each plague epidemic burial site is an adaptive response to the challenge of an abnormal death management situation, we will discuss the excavations that we performed in Venice, specifically on the island of Lazzaretto Vecchio (Fig. 3). All of the aforementioned sites are places that reveal the epidemic character of funeral management. During each plague epidemic, the communities had to develop places where they could treat the patients and where they could bury the victims. When new epidemics arrived years later, one or more new sites had to be utilized. The island of Lazzaretto Vecchio, which is situated in the middle of the lagoon in Venice, is an exception to this

FIGURE 3 Detailed view of a burial site during the excavation of Lazzaretto Vecchio at the Venice site (copyright Michel Signoli).

rule (10). From the end of the 14th century until the last epidemic in Venice in 1630, all the plague victims were buried on the island of Lazzaretto Vecchio. This is the only known example of the same place being utilized over several centuries for the sanitary funeral management of plague victims.

Although the funeral procedures discovered at burial sites provide a wealth of information, they are not the only criteria used to associate funeral sites with plague epidemics. The corpses observed during archaeological excavations and their comparisons with historical data allow researchers to define demographic profiles that are specific to plague, which differ from those for any other context.

DETAILS OF THE DEMOGRAPHIC REDUCTION AND THE BURIED VICTIMS DURING THE PLAGUE

A measurement of the importance and the details regarding the demographic reduction of human communities due to a lack of any effective measure to treat the plague is a fundamental subject that represents a crossing point for historical and osteo-archaeological sources. We will discuss this topic with regard to the well-documented case of Martigues during the plague epidemic of 1720-1721.

During the seven and one-half months of the epidemic, Martigues lost 2,134 inhabitants, which was approximately one-third of its total population, estimated as 6,069 people at that time (11). An examination of archived documents allowed us to restore local historical and demographic data and to build mortality patterns to represent the periods prior to, during, and following the epidemic (12). At the same time, both funeral sites were locally excavated. These sites provided a total sample of 250 skeletons, which were adequately preserved to allow an analysis of different biological parameters and to supply material for a paleodemographic study.

The plague mortality pattern established for Martigues between November of 1720 and June of 1721 shows that the risk for dying varied slightly with age. Across the population, the plague epidemic reduced the size of every age group, including those at the birth and fertility levels (i.e., births lost due to deaths in individuals at the age of procreation) over this time period. Moreover, the plague epidemic influenced the age structure of the adult population, although the composition of the age groups for the adult population did not differ greatly prior to and following the plague episode. Yet, the life force of the community showed considerable decline as more than one-third of the population died during the epidemic (Fig. 4). Given that the plague affected each age group similarly, an examination of the population according to 10-year age groups for the paleo-demographic sample should reflect the age structure of the living adult population on the eve of the epidemic (Fig. 5).

If the osteo-archaeological series is adequate, the plague funeral group should differ from that of a cemetery functioning during a period of normal mortality, as the group should appear consequently. It is a rare case in which the composition of the living population during a given time period can be precisely defined based on an examination of excavated skeletons. This account is consistent with the results of a study of the archaeological population of the graves of Martigues (3). This paleo-demographic study leads us to question whether it is possible to apply a comparable approach to other chronological contexts associated with the plague epidemic, of which the available documentation is limited to osteo-archaeological data. In other words, is it possible to find the structure and the number of individuals by age group for the adult population of a community from the distribution of skeletons by using the probability pattern established for Martigues in 1720-1721? The answer to this question is positive if one can provide evidence *a priori* supporting the hypothesis that the characteristics of the plague mortality, such as those discussed with regard to Martigues, are identical across the various epidemic crises of the second pandemic. If so, the "plague model" of Martigues can be applied to other paleo-demographic samples from plague funeral contexts.

There is a natural temptation to compare the mortality model established for the

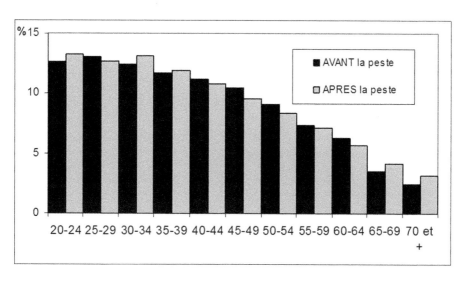

FIGURE 4 Distributions by age group (gender mixed) of the adult population before (November of 1720) and following (June of 1721) the plague episode. A nonparametric test (Mann-Whitney U test) revealed that there were no significant differences at the threshold of 0.01 between the two distributions.

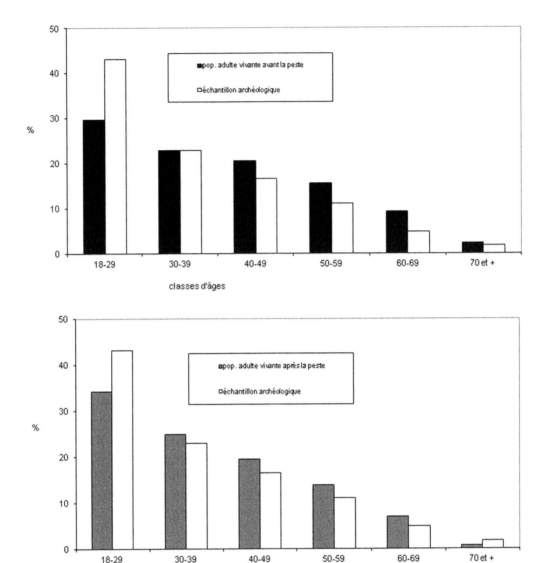

FIGURE 5 (Top) Proportional distributions of the living adult population by age group before the epidemic (November of 1720) and for the archaeological sample. The two series did not differ from each other at the threshold of 0.01. (Bottom) Proportional distributions of the living adult population by age group following the epidemic (June of 1721) and for the archaeological sample. The two series did not differ significantly from each other at the threshold of 0.01 (nonparametric Mann-Whitney U test).

plague period of 1720-1721 with those for other crises that resulted in abnormally high death rates from the beginning of the 18th century. By using the well-documented case in Martigues, it is possible to compare the distribution of deaths due to the plague of 1720-1721 according to 10-year age groups with the distribution of the deaths corresponding to an epidemic of smallpox in 1705 and with the distribution of deaths due to the wheat production crisis during the Great Frost in 1709-1710 (Fig. 6). It is important

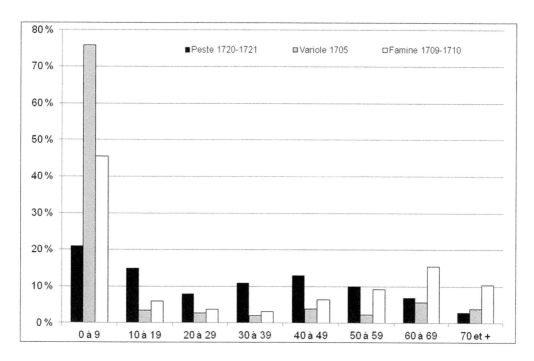

FIGURE 6 Comparative analysis of the resulting percentages (%) of demographic reduction for each ten-year age group due to the smallpox epidemic of 1705, the wheat production crisis during Great Frost in 1709-1710, and the plague epidemic of 1720-1721 in Martigues.

to note that these three crises of mortality did not result in similar quantitative types of demographic reduction. The crisis of 1705 killed 580 individuals (9.5% of the population of Martigues), whereas the Great Frost killed 1,468 individuals (24.2% of the population). In both of these scenarios, there is a decreased level of devastation compared with the plague that killed 2,134 individuals, which was more than one-third of the population (3).

The crisis of 1705 affected mainly young children, as 75% of the victims of this crisis were children younger than 10 years of age. The crisis of 1709-1710 had a bimodal distribution, with the greatest number of casualties evident in children younger than 10 years of age (45% of deaths) and adults older than 50 years of age (35% of deaths). The plague episode of 1720-1721 had a different profile in that all of the age groups were affected by the mortality rate in a manner proportional to their pre-epidemic representations.

Interestingly, research examining other provincial communities from the 18th century indicates that there are similar results regarding the demographic reduction due to smallpox and the wheat production crisis of 1709-1710. Only the plague generated differences in the decreasing population rates, which influenced approximately 60% of the population of certain communities, such as Berre, Valletta, and Néoules (13).

CITATION

Tzortzis S, Signoli M. 2016. Characterization of the funeral groups associated with plague epidemics. Microbiol Spectrum 4(4): PoH-0011-2015.

REFERENCES

1. **Signoli M.** 2006. *Étude anthropologique de crises démographiques en contexte épidémique.*

Aspects paléo- et biodémographiques de la peste en Provence. BAR International Series 1515, Oxford, UK.

2. **Paone F, Mellinand P, Parent F.** 2009. *Esplanade Major*, p 138–139. Bilan Scientifique Régional PACA 2008, Direction Régionale des Affaires Culturelles, Service Régional de l'Archéologie, Marseille, France.

3. **Tzortzis S.** 2009. Archives biologiques et archives historiques. Une approche anthropologique de l'épidémie de peste de 1720-1721 à Martigues (Bouches-du-Rhône, France). Thèse de doctorat d'anthropologie biologique. Université de la Méditerranée, Faculté de Médecine, Marseille, France.

4. **Tzortzis S, Signoli M.** 2009. Les tranchées des Capucins de Ferrières (Martigues, Bouches-du-Rhône, France). Un charnier de l'épidémie de peste de 1720-1722 en Provence. *C R Palevol* **8:**749–760.

5. **Bizot B, Castex D, Reynaud P, Signoli M.** 2005. *La saison d'une peste (avril-septembre 1590). Le cimetière des Fédons à Lambesc (Bouches-du-Rhône).* CNRS Éditions, Paris, France.

6. **Signoli M, Tzortzis S, Bizot B, Ardagna Y, Rigeade C, Seguy I.** 2007. Discovery of a 17th century plague cemetery (Puy-Saint-Pierre, Hautes-Alpes, France), p 115–119. *In* Signoli M, Cheve D, Adalian P, Boëtsch G, Dutour O (ed), *La peste: entre épidémies et sociétés. Actes du colloque ICEPID 4, 23-26 juillet 2001, Marseille.* Firenze University Press, Florence, Italy.

7. **Accoto J, Bello S, Bouttevin C, Castex D, Duday H, Dutour O, Moreau N, Pannuel M, Reynaud P, Signoli M.** 2005. Des données archéologiques et anthropologiques aux interprétations, p 37–62. *In* Bizot B, Castex D, Reynaud P, Signoli M (ed), *La saison d'une peste (avril-septembre 1590). Le cimetière des Fédons à Lambesc (Bouches-du-Rhône).* CNRS Éditions, Paris, France.

8. **Drancourt M, Aboudharam G, Signoli M, Dutour O, Raoult D.** 1998. Detection of 400-year-old Yersinia pestis DNA in human dental pulp: an approach to the diagnosis of ancient septicemia. *Proc Natl Acad Sci U S A* **95:**12637–12640.

9. **Bianucci R, Rahalison L, Rabino Massa E, Peluso A, Ferroglio E, Signoli M.** 2008. Technical note: a rapid diagnostic test detects plague in ancient human remains: an example of the interaction between archeological and biological approaches (Southeastern France, 16th-18th centuries). *Am J Phys Anthropol* **136:**361–367.

10. **Signoli M, Gambaro L, Rigeade C, Drusini A.** 2009. Les fouilles du Lazzareto Vecchio (Venise, Italie), p 333–346. *In* Buchet L, Rigeade C, Séguy I, Signoli M (ed), *Vers une anthropologie des catastrophes, Actes des 9e journées d'anthropologie de Valbonne.* Éditions INED/APDCA, Paris, France.

11. **Séguy I, Pennec S, Tzortzis S, Signoli M.** 2006. Modélisation de l'impact de la peste à travers l'exemple de Martigues (Bouches-du-Rhône), p 323–331 *In* Buchet L, Dauphin C, Seguy I (ed), *La paléodémographie. Mémoire d'os, mémoire d'hommes. Actes des 8e journées anthropologiques de Valbonne.* APDCA, Antibes, France.

12. **Séguy I, Signoli M, Tzortzis S.** 2007. Caractérisation des crises démographiques en basse Provence (1650-1725), p 197–204. *In* Signoli M, Cheve D, Adalian P, Boetsch G, Dutour O (ed), *Peste: entre épidémies et sociétés. Actes du colloque international ICEPID 4, Marseille, 23-26 juillet 2001.* Firenze University Press, Florence, Italy.

13. **Biraben JN.** 1975. *Les hommes et la peste en France et dans les pays européens et méditerranéens.* EHESS, Centre de Recherches Historiques, Mouton, Paris, France.

Paleogenetics and Past Infections: the Two Faces of the Coin of Human Immune Evolution

3

LAURENT ABI-RACHED[1] and DIDIER RAOULT[1]

PALEOGENETICS OF EXTINCT POPULATIONS

With the advent of next-generation sequencing, the field of paleogenetics has considerably expanded over the past few years, making investigations that were once considered impossible a reality. A milestone in paleogenetics was reached in the year 2010, which saw for the first time the reconstruction of the nuclear genome of ancient humans who lived thousands of years ago. The genomes characterized that year covered both modern humans, with the approximately 4,000-year-old Saqqaq man (1), and archaic humans, with the approximately 38,000-year-old Neanderthals from Croatia (2) and a more than 30,000-year-old Denisovan individual discovered in Siberia (3). These studies notably uncovered a migration of modern humans from Siberia into the New World some 5,500 years ago (1) and a new group of archaic humans who lived in Siberia, the Denisovans (3, 4). With these molecular data, it was also possible to refine the separation between modern and archaic humans to between 272,000 and 435,000 years ago, with a genetic divergence time of 734,000 to 1,087,000 years ago (2). Consistent with such an ancient genetic divergence between modern and archaic humans, for the human endogenous

[1]Aix Marseille Université, URMITE, UMR CNRS 7278, IRD 198, INSERM 1095, Faculté de Médecine, Institut Hospitalo-Universitaire Méditerranée-Infection, Marseille, France.
Paleomicrobiology of Humans
Edited by Michel Drancourt and Didier Raoult
© 2016 American Society for Microbiology, Washington, DC
doi:10.1128/microbiolspec.PoH-0018-2015

retrovirus (HERV) that is thought to have promoted the development of the placenta in mammals (5), archaic individuals have six HERV-K proviruses (three common to Denisovans and Neanderthals, two specific to Neanderthals, and one specific to the Denisovan individual) that are absent in a 402-genome set of modern-day individuals (6). Following these early characterizations of draft genomes of archaic humans, efforts focused on reconstructing high-quality genome sequences, and this was achieved for the original Denisovan genome (7) and for the genome of a Neanderthal individual who lived in Altai (Fig. 1) (8).

Comparison of these genomes of archaic humans with those of modern humans revealed clear evidence of admixture between the two groups; Neanderthals contributed 1.5% to 2.1% to modern Eurasian genomes, and Denisovans contributed 3% to 6% to modern Melanesian genomes and to approximately 0.2% of the genomes of Asians (8). Admixture was also detected between archaic groups; 0.5% of the Denisovan genome was of Neanderthal origin and another 0.5% to 8% originates from of an unknown hominin (8). Therefore, it is now clear that ancient humans interbred with morphologically extinct Denisovan and Neanderthal

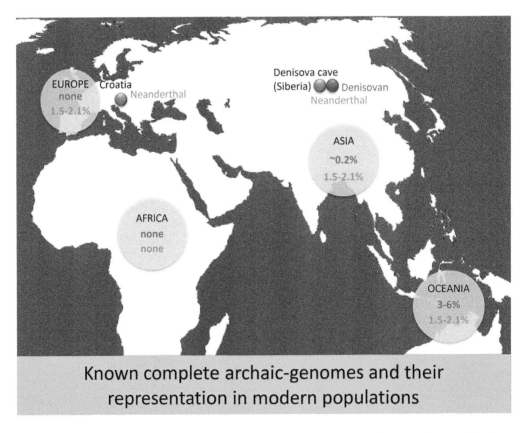

Known complete archaic-genomes and their representation in modern populations

FIGURE 1 **Known complete archaicgenomes and their representation in modern populations. The three complete nuclear genomes of archaic humans that have been reconstructed to date are indicated by red (Denisovan) or blue (Neanderthal) circles at the geographical location where the samples used for these reconstructions were uncovered. The impact of archaic humans on the genomes of modern humans—as measured by the average proportion of the genome that is of archaic origin—is given for four regions of the world in gray circles. Red/blue: proportions of the genome that are of Denisovan (red) or Neanderthal (blue) origin (8).**

individuals (9), and that modern-day populations are not descended from a single evolutionary branch but are instead a mosaic of anatomically modern humans and Neanderthal, Denisovan, and other archaic humans (10).

Yet, the nature of the encounters that produced the first hybrids is still an outstanding question. A recent study suggests that the reduced level of Neanderthal ancestry in Eurasian X chromosomes is due to male hybrid sterility (11). However, X chromosomes are always transmitted by mothers, whereas fathers transmit them only half of the time. Therefore, the reduced level of Neanderthal ancestry in Eurasian X chromosomes could be explained by encounters between archaic men and modern women having been far more common than the opposite. This "Neanderthal fathers" model would also explain why modern humans lack Neanderthal mitochondrial DNA (mtDNA) because mtDNA is transmitted only by mothers. Consistent with this, the chromosome X-to-autosomes ratio of archaic ancestry in the least functionally important regions (categories '4-5' in Fig. 2 of reference 11), which are more likely to evolve under neutrality, is biased toward the 50% level (mating having involved only archaic men). This bias, which is slightly more marked in Europeans (60% to 62%) than in Asians (64% to 75%), thus suggests that the Neanderthal ancestors of modern Eurasians were mostly men. According to Ockham's razor principle, this suggests that the observed pattern of Neanderthal ancestry in modern Eurasian populations is linked to behaviors such as mate selection or even rape.

While these breakthroughs in paleogenetics have had an immediate impact on our understanding of human evolution, they also make it possible to derive more precisely the evolution of rapidly evolving gene families, such as those of the immune system. This in turn can help us to understand past adaptations against infectious diseases and in particular the role past epidemics have played in shaping modern immune systems. Here, we briefly tackle this topic by reviewing and discussing how human paleogenetics can contribute to an improvement of our understanding of the interplay between pathogens and their host populations.

PALEOIMMUNOLOGY OF PAST POPULATIONS

Paleogenetics allows direct investigations of the genetics of ancient and extinct populations. This is critical in immunology because it makes it possible to reconstruct precisely the evolution of the rapidly evolving gene families of the immune system. In particular, the availability of archaic genome sequences allows an unprecedented exploration of the role admixture with extinct populations has played in this evolution and a consideration of the possibility of adaptive introgression.

One of the first such studies of the possible role that admixture with extinct populations has played in the evolution of the immune system started as an investigation of HLA-B*73, a divergent lineage of the HLA-B locus that has an unusually low level of genetic diversity in modern populations and that is more prevalent in West Asia and rare in other regions of the world (12). These characteristics are consistent with an archaic origin, a model supported by population simulations and a result that prompted a more general investigation of the impact of archaic admixture on modern HLA class I content. *In silico* HLA class I typing of the Denisovan and Neanderthal genomes hence revealed the presence of alleles that were common in modern human populations. That these shared alleles were present on long and shared haplotypes indicated they originated from recent admixture, and putative archaic ancestry for the *HLA-A* locus was estimated to be more than 50% in Europe, more than 70% in Asia, and more than 95% for some populations in Oceania (12). More recently, it was also shown that modern-day populations have inherited Denisovan and Neanderthal versions of the Toll-like receptors (TLRs) implicated in the immune response against pathogens, and that these variants of archaic origin are expressed at higher levels than the non-introgressed variants (13). These studies thus indicate that even though admixture with archaic humans contributed to only a small fraction of the modern genomes (Fig. 1), this has had a profound impact on systems like the immune system.

Besides the tremendous efforts to derive paleoimmunology from paleogenetics, another approach to paleoimmunology is to retrieve ancient immune system molecules, chiefly immunoglobulins in ancient individuals. Indeed, a few seminal works showed that it was possible to detect proteins in ancient mammalian specimens and even to reconstitute the protein-based phylogeny of some extinct species (14, 15). Yet, in these studies, limited attention was given to the conservation of immunoglobulin fragments as pieces of the immunological framework in ancient individuals (16). Instead, a set of 13 inflammatory proteins was, for example, identified in a well-preserved 500-year-old Inca mummy of a young girl (15 years old at death) with probable lung infection (16). The study of the immune reactivity of ancient humans is, however, possible, as illustrated by the reactivity of bone-extracted IgG immunoglobulins against the syphilis agent *Treponema pallidum* in one 18th century individual with a diagnosis of syphilis on the basis of the PCR amplification and sequencing of specific markers (17). Although these examples are based on individuals who were buried relatively recently, they demonstrate that it is possible to recover inflammatory proteins and immunoglobulins in ancient individuals. We propose that this search for immune system proteins in selected specimens, such as preserved dental pulp, could complement genome-based analyses to depict the functionality of the immune system in some ancient populations.

SELECTIVE PRESSURE OF PAST EPIDEMICS

The genetic variation of the immune system affects individual resistance or susceptibility to infectious agents, and selection for the resistance traits can reshape immune diversity in populations. The malaria agent *Plasmodium falciparum* is, for example, suspected to contribute to the co-evolution and selection of HLA class I and killer cell immunoglobulin-like receptors (KIRs) in sub-Saharan populations (18). As paleogenetics characterizes cases of adaptive introgression that occurred hundreds or thousands of years ago, the link with past epidemics can be difficult to establish, but phenotypes vis-à-vis of modern infectious agents can provide some clues. For example, the HLA-A*11 allotype that is of Denisovan origin and has a high frequency in Asia and Oceania led to a unique Epstein-Barr virus evolution in those populations with a high frequency of the allotype, suggesting that HLA-A*11 can exert strong selective pressures on viruses (19). Similarly, the TLR variants inherited from archaic humans are associated with a reduced risk for *Helicobacter pylori* infection, hence providing a link to bacterial infections (13).

The same approaches led to the hypothesis that the 32-bp deletion in the C-C chemokine receptor type 5 (CCR5), nowadays associated with resistance to HIV infection (20), was selected by plague epidemics (21). HIV became a human pathogen only in recent history, but the emergence of the protective variant, CCR5Δ32, is much more ancient, with estimates ranging from 700 years ago based on coalescent theory (21) to 2,900 years ago based on molecular studies (22). CCR5Δ32 is also restricted to Europe, with a north-to-south gradient of allele frequency of 16% to 5% (23). The age of the variant and its high prevalence in Europe are thus consistent with positive selection linked to an earlier human pathogen, and the one that caused the medieval plague (Black Death), which likely killed as much as 30%

to 50% of the European population (24), was suggested as a candidate (21). However, experimental models aimed at confirming this hypothesis yielded conflicting results (25–27), and alternative hypotheses have been proposed (28). Our analysis for the presence of CCR5Δ32 in 53 individuals from a 1722 plague mass grave in southern France revealed that 40 of the 53 individuals (75.5%) were wild-type, 12 (22.6%) were heterozygous, and 1 was homozygous for CCR5Δ32. Positive PCR amplification of the *Y. pestis* glp D gene also confirmed that this individual was indeed infected by *Y. pestis* (Orientalis-like genotype) at the time of death. These data show that individuals homozygous for CCR5Δ32 are susceptible to infection by *Y. pestis*, demonstrating that CCR5Δ32 does not provide complete protection against the plague, as is the case against some HIV strains.

PROVISIONAL CONCLUSIONS

We are entering an exciting period in which paleogenetics and paleomicrobiology data can be integrated to generate a clearer picture of how the immune system of modern populations were shaped in each region of the world and the roles that admixture and epidemics have played in such evolutions.

CITATION

Abi-Rached L, Raoult D. 2016. Paleogenetics and past infections: the two faces of the coin of human immune evolution. Microbiol Spectrum 4(3):PoH-0018-2015.

REFERENCES

1. Rasmussen M, Li Y, Lindgreen S, Pedersen JS, Albrechtsen A, Moltke I, Metspalu M, Metspalu E, Kivisild T, Gupta R, Bertalan M, Nielsen K, Gilbert MT, Wang Y, Raghavan M, Campos PF, Kamp HM, Wilson AS, Gledhill A, Tridico S, Bunce M, Lorenzen ED, Binladen J, Guo X, Zhao J, Zhang X, Zhang H, Li Z,

Chen M, Orlando L, Kristiansen K, Bak M, Tommerup N, Bendixen C, Pierre TL, Gronnow B, Meldgaard M, Andreasen C, Fedorova SA, Osipova LP, Higham TF, Ramsey CB, Hansen TV, Nielsen FC, Crawford MH, Brunak S, Sicheritz-Ponten T, Villems R, Nielsen R, Krogh A, Wang J, Willerslev E. 2010. Ancient human genome sequence of an extinct Palaeo-Eskimo. *Nature* **463**:757–762.

2. Green RE, Krause J, Briggs AW, Maricic T, Stenzel U, Kricher M, Patterson N, Heng L, Zhai W, Fritz MH, Hansen NF, Durand EY, Malaspinas A, Jensen JD, Marques-Bonet T, Alkan C, Prüger K, Meyer M, Burbano HA, Good JM, Schultz R, Aximu-Petri A, Butthof A, Höber B, Höffner B, Siegemund M, Weihmann A, Nusbaum C, Lander ES, Russ C, Novod N, Affourtit J, Egholm M, Verna C, Rudan P, Brajkovic D, Kucan Z, Gusic I, Doronichev VB, Golovanova LV, Lalueza-Fox C, Rasilla M, Fortea J, Rosas A, Schmitz RW, Johnson PLF, Eichler EE, Falush D, Birney E, Mullikin JC, Slatkin M, Nielsen R, Kelso J, Lachmann M, Reich D, Pääbo S. 2010. A draft sequence of the Neandertal genome. *Science* **328**:710–722.

3. Reich D, Green RE, Kircher M, Krause J, Patterson N, Durand EY, Viola B, Briggs AW, Stenzel U, Johnson PLF, Maricic T, Good JM, Marques-Bonet T, Alkan C, Fu Q, Mallick S, Li H, Meyer M, Eichler EE, Stoneking M, Richards M, Talamo S, Shunkov MV, Derevianko AP, Hublin J, Kelso J, Slatkin M, Pääbo S. 2010. Genetic history of an archaic hominin group from Denisova Cave in Siberia. *Nature* **468**:1053–1060.

4. Krause J, Fu Q, Good JM, Viola B, Shunkov MV, Derevianko AP, Pääbo S. 2010. The complete mitochondrial DNA genome of an unknown hominin from southern Siberia. *Nature* **464**:894–897.

5. Chuong EB. 2013. Retroviruses facilitate the rapid evolution of the mammalian placenta. *Bioessays* **35**:853–861.

6. Lee A, Huntley D, Aiewsakun P, Kanda RK, Lynn C, Tristem M. 2014. Novel Denisovan and Neanderthal retroviruses. *J Virol* **88**:12907–12909.

7. Meyer M, Kircher M, Gansauge MT, Li H, Racimo F, Mallick S, Schraiber JG, Jay F, Prüfer K, de Filippo C, Sudmant PH, Alkan C, Fu Q, Do R, Rohland N, Tandon A, Siebauer M, Green RE, Bryc K, Briggs AW, Stenzel U, Dabney J, Shendure J, Kitzman J, Hammer MF, Shunkov MV, Derevianko AP, Patterson N, Andrés AM, Eichler EE, Slatkin M, Reich D, Kelso J, Pääbo S. 2012. A high-coverage genome sequence from an archaic Denisovan individual. *Science* **338**:222–226.

8. Prüfer K, Racimo F, Patterson N, Jay F, Sankararaman S, Sawyer S, Heinze A, Renaud G, Sudmant PH, Filippo C, Li H, Mallick S, Dannemann M, Fu Q, Kircher M, Kuhlwilm M, Lachmann M, Meyer M, Ongyerth M, Siebauer M, Theunert C, Tandon A, Moorjani P, Pickrell J, Mullikin JC, Vohr SH, Green RE, Hellman I, Johnson PLF, Blanche H, Cann H, Kitzman JO, Shendure J, Eichler EE, Lein ES, Bakken TE, Golovanova LV, Doronichev VB, Shunkov MV, Derevianko AP, Viola B, Slatkin M, Reich D, Kelso J, Pääbo S. 2014. The complete genome sequence of a Neanderthal from the Altai Mountains. *Nature* **505**:43–49.

9. Callaway E. 2015. Neanderthals had outsize effect on human biology. *Nature* **523**:512–513.

10. Birney E, Pritchard JK. 2014. Archaic humans four makes a party. *Nature* **505**:32–34.

11. Sankararaman S,Mallick S, Dannemann M, Prüfer K, Kelso J, Pääbo S, Patterson N, Reich D. 2014. The genomic landscape of Neanderthal ancestry in present-day humans. *Nature* **507**:354–357.

12. Abi-Rached L, Jobin MJ, Kulkarni S, McWhinnie A, Dalva K, Gragert L, Babrzadeh F, Gharizadeh B, Luo M, Plummer FA, Kimani J, Carrington M, Middleton D, Rajalingam R, Beksac M, Marsh SG, Maiers M, Guethlein LA, Tavoularis S, Little AM, Green RE, Norman PJ, Parham P. 2011. The shaping of modern human immune systems by multiregional admixture with archaic humans. *Science* **334**:89–94.

13. Dannemann M, Andres AM, Kelso J. 2016. Introgression of Neandertal- and Denisovan-like haplotypes contributes to adaptive variation in human Toll-like receptors. *Am J Hum Genet* **98**:22–33.

14. Asara JM, Schweitzer MH, Freimark LM, Phillips M, Cantley LC.. 2007. Protein sequences from mastodon and *Tyrannosaurus rex* revealed by mass spectrometry. *Science* **316**:280–285.

15. Schweitzer MH, Zheng W, Organ CL, Avci R, Suo Z, Freimark LM, Lebleu VS, Duncan MB, Vander Heiden MG, Neveu JM, Lane WS, Cottrell JS, Horner JR, Cantley LC, Kalluri R, Asara JM. 2009. Biomolecular characterization and protein sequences of the Campanian hadrosaur *B. canadensis*. *Science* **324**:626–631.

16. Corthals A, Koller A, Martin DW, Rieger R, Chen EI, Bernaski M, Recagno G, Dávalos LM. 2012. Detecting the immune system response of a 500 year-old Inca mummy. *PLoS ONE* **7**:e41244. doi:10.1371/journal.pone.0041244.

17. Kolman CJ, Centurion-Lara A, Lukehart SA, Owsley DW, Tuross N. 1999. Identification of *Treponema pallidum* subspecies *pallidum* in a 200-year-old skeletal specimen. *J Infect Dis* **180**:2060–2063.

18. Norman PJ, Hollenbach JA, Nemat-Gorgani N, Guethlein LA, Hilton HG, Pando MJ, Koram KA, Riley EM, Abi-Rached L, Parham P. 2013. Co-evolution of human leukocyte antigen (HLA) class I ligands with killer-cell immunoglobulin-like receptors (KIR) in a genetically diverse population of sub-Saharan Africans. *PLoS Genet* **9**:e1003938. doi:10.1371/journal.pgen.1003938.

19. Campos-Lima PO, Gavioli R, Zhang QJ, Wallace LE, Dolcetti R, Rowe M, Rickinson AB, Masucci MG. 1993. HLA-A11 epitope loss isolates of Epstein-Barr virus from a highly A11 + population. *Science* **260**:98–100.

20. Allers K, Schneider T. 2015. CCR5Δ32 mutation and HIV infection: basis for curative HIV therapy. *Curr Opin Virol* **14**:24–29.

21. Stephens JC, Reich DE, Goldstein DB, Shin HD, Smith MW, Carrington M, Winkler C, Huttley GA, Allikmets R, Schriml L, Gerrard B, Malasky M, Ramos MD, Morlot S, Tzetis M, Oddoux C, di Giovine FS, Nasioulas G, Chandler D, Aseev M, Hanson M, Kalaydjieva L, Glavac D, Gasparini P, Kanavakis E, Claustres M, Kambouris M, Ostrer H,Duff G, Baranov V, Sibul H, Metspalu A, Goldman D, Martin N, Duffy D, Schmidtke J, Estivill X, O'Brien SJ, Dean M. 1998. Dating the origin of the CCR5-Delta32 AIDS-resistance allele by the coalescence of haplotypes. *Am J Hum Genet* **62**:1507–1515.

22. Hummel S, Schmidt D, Kremeyer B, Herrmann B, Oppermann M. 2005. Detection of the CCR5-Delta32 HIV resistance gene in Bronze Age skeletons. *Genes Immun* **6**:371–374.

23. Lucotte G. 2002. Frequencies of 32 base pair deletion of the (Delta 32) allele of the CCR5 HIV-1 co-receptor gene in Caucasians: a comparative analysis. *Infect Genet Evol* **1**:201–205.

24. Signoli M. 2012. Reflections on crisis burials related to past plague epidemics. *Clin Microbiol Infect* **18**218–223.

25. Mecsas J, Franklin G, Kuziel WA, Brubaker RR, Falkow S, Mosier DE. 2004. Evolutionary genetics: CCR5 mutation and plague protection. *Nature* **427**:606.

26. Elvin SJ, Elvin SJ, Williamson ED, Scott JC, Smith JN, Pérez De Lema G, Chilla S, Clapham P, Pfeffer K, Schlöndorff D, Luckow B. 2004. Evolutionary genetics: ambiguous role of CCR5 in *Y. pestis* infection. *Nature* **430**:417.

27. **Styer KL, Click EM, Hopkins GW, Frothingham R, Aballay A.** 2007. Study of the role of CCR5 in a mouse model of intranasal challenge with *Yersinia pestis. Microbes Infect* **9:**1135–1138.

28. **Galvani AP, Slatkin M.** 2003. Evaluating plague and smallpox as historical selective pressures for the CCR5-Delta 32 HIV-resistance allele. *Proc Natl Acad Sci U S A* **100:**15276–15279.

A Personal View of How Paleomicrobiology Aids Our Understanding of the Role of Lice in Plague Pandemics

4

DIDIER RAOULT[1]

HISTORICAL DESCRIPTION

As early researchers in paleomicrobiology, we have used its techniques to study the plague, generating controversy after the publication of our initial results (1). The controversy has allowed us to respond and propose new approaches to understanding plague pandemics. Our conclusion is that the plague was related to outbreaks of lice, not to rat fleas, because fleas cannot explain epidemics of this magnitude (1).

The Black Death first appeared in the 14th century in the Near East and very quickly spread to Europe (1). It was well described by Guy de Chauliac, a physician to the pope in Avignon, who described a bubonic epidemic disease. We know of no other epidemic diseases that are associated with buboes—i.e., suppurating glands in the groin (*bubo* is the Latin word for groin). Many paintings show individuals with buboes during the plague, in Renaissance art, which leaves no doubt about the origin of this disease. For example, there are several representations of Saint Sebastian with buboes of the plague as Roman arrows strike him during his ordeal. Procopius, during the reign of Justinian in the 6th century, gave an extremely detailed description of the symptoms of the plague (1). Thus, for a physician who

[1]URMITE, UM63, CNRS 7278, IRD 198, Insert 1095, Aix Marseille University, Marseille, France.
Paleomicrobiology of Humans
Edited by Michel Drancourt and Didier Raoult
© 2016 American Society for Microbiology, Washington, DC
doi:10.1128/microbiolspec.PoH-0001-2014

reads the original texts, the origin of the two pandemics—i.e., Justinian's plague and the medieval plague, there is no doubt about their origin. There is a frequent misrepresentation in descriptions of the plagues of the Middle Ages, which have been called the Black Plague or the Black Death because of the serious nature of the disease, not because of a particular clinical manifestation (2).

TWO DESCRIPTIONS OF THE DISEASE CYCLE IN THE 20TH CENTURY

In the 20th century, most of the studies were performed by French investigators and published in French. Yersin found the microbe responsible for the disease (2); it was first called *Pasteurella pestis* and then *Yersinia pestis*, in honor of Yersin. It was discovered that the rat flea was an important vector for carrying the microbe (Simond) (3–5). An elegant experiment was performed showing that transmission from one rat to another was possible only when infected fleas jumped up from a lower cage to a higher cage. It was thus concluded and widely accepted that rat fleas transmitted the disease. The plague cycle was particularly well described by Baltazard in Iran (2, 6), who showed that when outbreaks of epizootic plague occurred in wild rodents, fleas that infected the wild rodents would then bite humans, resulting in sporadic cases of plague in humans. However, pandemics of plague remained rare, even though sporadic cases were still common, similar to what is seen today (7). This explanation of pandemics is problematic because it is not likely that a whole population would be infected only by rat flea bites (8–10). Baltazard identified *Y. pestis* in lice during an epidemic in Morocco (11). The assumption generally accepted in the French school and developed by Mollaret (3, 12) was that the pandemic phase was linked to human ectoparasites, especially

the common flea, in ancient times (*Pulex irritans*), not to the rat flea. The role of human transmission via pneumonia was also examined, but it is now known that the risk for transmission via this route is extremely low (13). Nevertheless, the number of species of fleas capable of transmitting plague is extremely large (3). Unfortunately, very little work has been done on the plague and its transmission for decades, which has led to a simplification of the plague epidemic cycle between anthropophilic rats (*Rattus rattus* and *Rattus norvegicus*) and their fleas; the fleas then bite humans, and person-to-person transmission occurs through coughing. This was the state of research when we started our paleomicrobiology work.

BIRTH OF PALEOMICROBIOLOGY

Paleomicrobiology was partially born in my laboratory in 1996, when a tunnel was dug in Marseille. This excavation uncovered an ancient grave that was clearly identified as a mass grave from the Great Plague of Marseille (1720), which killed a large part of the population and was a resurgence of the great plague of the Middle Ages (14). Olivier Dutour, who conducted the investigation of this mass grave, had already done preliminary paleomicrobiological work with us when he asked us to try (without success) to identify the bacterium responsible for syphilis (*Treponema pallidum*) in European bones with lesions from the pre-Columbian period (to challenge the assumption that syphilis was brought to Europe by sailors who voyaged with Columbus).

When investigating this mass grave, Dr. Dutour came to my laboratory with a carton of bones and skulls from people who had died of plague and asked what could be done to confirm that it was indeed the plague that had killed them. By chance, there was a dentist in my laboratory, Gérard

Aboudharam, doing his master's degree work with me. He had very little availability and could not work in the laboratory more than two afternoons per week, so that it was difficult to find an area where we could use his skills and still allow him to make a significant contribution. At this time, however, I asked him to look inside the dental pulp of the plague victims for the remains of *Y. pestis*, joking that "you are going to work on dental plague!" Teeth are made of dentin, and inside is a highly vascularized small organ called the dental pulp. We therefore thought that if a microorganism (i.e., *Y. pestis*) had infected the blood when the patient died, we would be able to detect the bacterial DNA in the dental pulp. This would be similar to a sample of dried blood, which could also be used to obtain a DNA sample. We produced two experimental models with *Coxiella burnetii* (15, 16) in guinea pigs to show that the technique was efficient and that it was possible to find agents of bacteremia within the dental pulp.

We first examined the teeth of those who died in the 1720 plague in Marseilles, and by using PCR, we were able to identify, for the first time, *Y. pestis* in the dental pulp of these people (Fig. 1) (14). Along with Michel Drancourt and Gérard Aboudharam, I sent this work to *Proceedings of the National Academy of Sciences*, and I was fortunate to have as editor Nobel Prize winner Joshua Lederberg, who had defined the emerging diseases of the 20th century and who immediately saw the value of this work, which has since been cited more than 200 times. Following this publication, a British researcher, Dr. M. B. Prentice, asked to learn our technique, so we trained him in dental pulp extraction and the use of our primers for performing PCR. It should be noted that before this time, *Y. pestis* had never been amplified in our laboratory, and thus the chance of contamination was null. Furthermore, our laboratory is known for performing complex molecular tests, for

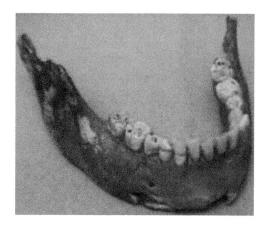

FIGURE 1 **Recovery of dental pulp from an ancient jaw. To view video go to http://www.asmscience. org/files/recovery_of_human_dental_pulp.mpg.**

which we receive thousands of samples from hospitals from around the world. Our work was subsequently validated by a second study in which I developed a new technique, "suicide PCR," which allowed us to avoid any risk of contamination (17). In this technique, we used a sequence target and primers only once to avoid generating amplicons that could contaminate later studies.

THE CONTROVERSY

A dual controversy arose from this publication, led on the one hand by C. J. Duncan (18, 19) and related to epidemiological and demographic considerations, and on the other by an English team that failed to reproduce our results (18, 20). Careful attention was paid to alternative hypotheses for explaining the great plague of the Middle Ages. The press got hold of these hypotheses and reported that the great plague of the Middle Ages had been hemorrhagic fever, most likely caused by the Ebola and Marburg viruses, and that we were ill prepared to prevent another such outbreak because we still do not have any weapons to fight these viruses (18). This work has been the subject of two best-selling books, yet the development perspective of the plague, saying that it was related to *Y. pestis*, undoubtedly challenges these works. Moreover, some of the scientific work based on these hypotheses were funded by the National Scientific Foundation (NSF), and for demographers, in fact, all meaning would be lost if the plague were merely due to *Y. pestis*. The authors of a best seller book claiming that black death was caused by hemorragic fever sent a letter to *Lancet Infectious Diseases* (21), and I responded by saying that the question of the plague was not an issue that arose, nor was it from the genome of the skeleton identified as Grand Duchess Anastasia of Russia and not descendants of people who escaped the Soviet massacre (22). It is interesting to note that two authors, M. B. Prentice and T. Gilbert, an anthropologist, responded that our work was wrong and most likely represented a contamination because they themselves had failed to reproduce it (20). Their work was published in the journal *Microbiology* and showed that they had not been able to use our technique and had used samples other than those we ourselves tested (20). Among the samples they tested, we shared a single lot of teeth; however, our laboratory also obtained negative results for these particular teeth. We proposed an exchange of teeth lots and that we would test theirs and they would test ours, with blind controls, but we never received any proposal (22). When we contacted them, Prentice said he no longer worked on this topic, and so it has never been scientifically determined whether we had contamination or they not were able to amplify DNA from *Y. pestis*. The basis of their technical controversy was the belief that the DNA was most likely too degraded for it to be possible to amplify sequences of this size. This theory, however, does not agree with our results. Indeed, this study influenced the British media (including the BBC) and Wikipedia!

THE CONFIRMATION

To confirm our work, I invented a technique, "suicide PCR," in which PCR is used only once, so that no amplicons can contaminate successive tests. This work was published in *Proceedings of the National Academy of Sciences* and was edited by Joshua Lederberg (17). Since then, other teams have worked on the same domain to test this hypothesis. We have successfully confirmed, for the first time, that Justinian's plague was related to *Y. pestis* by developing a new genotyping technique to obtain original sequences that could not result from contamination (23). For this demonstration, we developed a technique called multispacer typing (which was used for the first time in this disease) by taking the most variable noncoding regions, speculating that they would contain the most single-nucleotide position mutations (SNIPs). This was also the subject of controversy because the author who used microsatellites techniques challenged the validity of the search for SNIPs (which became the basis for genotyping), but this controversy fizzled (24) once many other teams had tested it (14). The first to confirm our work on Justinian's plague was a German team, and then other international teams also validated our work. Additionally, other techniques have emerged to sequence the complete genome of *Y. pestis* and have been the subject of several publications (23, 25–30). The most famous of these, published in *Nature*, used the dental pulp material that we described first, and the authors cited no other references on the topic (including ours, despite the fact that their paper was published 15 years after our initial paper in *Proceedings of the National Academy of Sciences*) (31). In some places, this is considered scientific misconduct! *Nature* rejected my letter of protest regarding this paper. Another paper published by that same team confirmed that dental pulp was superior to bone samples to recover ancient DNA, in contrast to the conclusions of Gilbert et al. (20, 25). The review of this work had not been done properly, and the authors had likely asked that we be excluded as reviewers because we would have immediately noticed that these authors did not properly cite references and that they had "rediscovered" something that was found over 15 years ago. This is a lesson that I teach—science is a fight in which people may be cheating (https://www.youtube.com/watch?v=QOc__QvYJ-E). Either way, there is little doubt that *Y. pestis* was the cause of Justinian's plague and the great plague of the Middle Ages. Gilbert initially raised questions about our work by writing that *Y. pestis* was most likely not the cause of the plague and that the teeth we used were most likely not the right place to look for the agent of the plague. Recently, an editorial appeared in *Lancet Infectious Diseases* (32) commenting on the results of Poinard (33), finally welcoming the idea that identification of the *Y. pestis* plague could be performed in samples that he had tested (and found negative) and by using dental pulp!

SIZE OF ANCIENT DNA

One of misleading hypothesis (20) was that ancient DNA was too degraded to be amplified. This was based only on theoretical analysis and neglected the fact that PCR can amplify the few nondegraded molecules. In fact, most of these propositions were based on theoretical models (34) and were not confirmed experimentally. It was found that eukaryotic DNA is degraded more rapidly than bacterial DNA (35). This is not surprising, as bacteria, including *Y. pestis*, can survive in a dormant state for years in the soil (36). An independent study (37) confirmed that bacterial DNA from ancient human dental pulp can be amplified by 16S rDNA universal primers, finally resolving the 15-year-old debate of the "too

degraded DNA" during which we had to fight to defend our work.

LOOKING FOR AN EPIDEMIOLOGICAL EXPLANATION

Ancient Studies

It was clear that the simplified cycle, used in the past by demographers to model the outbreak, was wrong. The cycle was summarized with rat, its fleas, and respiratory transmission (38). Obviously, this model lacked many parameters in order to explain a complex diffusion with attack rates of nearly 90%. We then began to think that the louse could be implicated in this outbreak. Indeed, people were likely covered in lice until recently, and the worst pandemics for which this vector has been identified were related to pandemic lice, including relapsing fever, typhus, and trench fever due to *Bartonella quintana* (39). Thus, having found Baltazard's work identifying *Y. pestis* in lice, it convinced us to test this hypothesis (11).

Experimental Model of Lice Infection

We had developed experimental models of infection with *Rickettsia prowazekii* (40), other *Rickettsia* species, *Borrelia* (41–44), and *B. quintana* (40), and therefore we decided to develop an experimental model of rabbits infected with *Y. pestis* lice as well. We showed that rabbits that had been fed on human lice and were infected with *Y. pestis* themselves infected new, naive lice that were now capable of transmitting *Y. pestis* to uninfected rabbit (1, 45). Thus, we had an experimental model demonstrating the ability of lice to transmit *Y. pestis* (1).

Co-infection with *Bartonella quintana*

Our work on paleomicrobiology with Michel Drancourt and Michel Signoli continued to grow, and we ended up accumulating a series of 2000 teeth to analyze (26). We organized a comprehensive program to systematically test all pathogens that could possibly be in the mass graves where the teeth originated, including anthrax, plague, *B. anthracis*, *Y. pestis*, *B. quintana*, *R. prowazekii*, and the smallpox virus (27). Interestingly, we could identify in a household the presence of *Y. pestis* and *B. quintana* (27). *B. quintana* is transmitted primarily by lice but may occasionally be transmitted by fleas from cats. The existence of a dual epidemic, caused by *B. quintana* and *Y. pestis*, reinforced the hypothesis that some of epidemics related to lice could be caused by both *Y. pestis* and *B. quintana*. In previous work on typhus, we found an epidemic of lice contemporaneously in Burundi (46) and in the great army of Napoleon (47). We tested the remains of soldiers who died after crossing the Berezina and found infections that were caused by either *B. quintana* or *R. prowazekii*. Thus, we demonstrated the existence of dual epidemics from a single vector.

Louse infection with *Yersinia pestis*

We started looking into different places to identify lice infected with *Y. pestis*. For several years, we tried to get lice from Madagascar, but it was through our connection with Renaud Piarroux in the eastern Congo that we obtained access to lice from an area where plague is endemic and renewed the work of Baltazard of finding *Y. pestis* DNA in a contemporary louse (48). Moreover, in this study, we found both *B. quintana* and *Y. pestis* in the lice we sampled.

CONCLUSION

Our paleomicrobiological study of plague due to *Y. pestis* forced us to completely reanalyze the epidemiology of pandemic plague. We were forced by the controversy surrounding our work (work that was later validated by all the teams that had tested our samples except one) to reanalyze the role of lice in these pandemics and to consider the likelihood that pandemics were linked to an outbreak of lice in the affected populations. Evidence that plague pandemics were transmitted by lice was as follows:

- *Y. pestis* found in body lice
- Experimental model of infection with human lice in the rabbit
- Paleomicrobiological evidence of epidemics involving both *Y. pestis* and *B. quintana* (a louse-transmitted disease causing trench fever)

This is similar, in fact, to what I saw in Africa during the civil war that raged in Rwanda and Burundi in the 1990s. Initially, I was asked by the World Health Organization (WHO) to examine a possible outbreak of typhus in the Goma refugee camp with nearly 800,000 refugees, all infected with head lice. No outbreak of *R. prowazekii* infection occurred at that time (49). Instead, these refugees likely had infection due to *B. quintana*, which is much more common but has not been tested at this time. However, shortly after my visit to Goma for WHO, in 1994, I was asked by a student from Burundi (who had studied in Marseille) to investigate a small epidemic of typhus in a prison in Burundi (50), and I then had the opportunity to go there to investigate a larger epidemic developing in the refugee camps (which I estimated to be nearly 100,000 cases [51]), during which an estimated 10,000 people died of typhus (46). My investigation of this contemporary epidemic allowed me to show, for the first time, a dual epidemic of diseases transmitted by lice: *B. quintana* and *R. prowazekii*. It is reasonable to assume that plague pandemics were associated with the presence of lice and that the infection of a patient carrying lice infected with *Y. pestis* allowed the dissemination and development of these pandemics. It should be noted that among the pandemics currently identified, respiratory and lice-related pandemics are among the most devastating in the history of mankind. Based on paleomicrobiology, the involvement of lice can be considered to be highly probable. In conclusion, our work and the controversy surrounding it encouraged us to research plague pandemic mechanisms and taught us that you can publish a paper in a famous scientific journal (*Nature*) without referring to frequently cited papers related to the same topic.

CITATION

Raoult D. 2016. A personal view of how paleomicrobiology aids our understanding of the role of lice in plague pandemics. Microbiol Spectrum 4(4):PoH-0001-2014.

REFERENCES

1. **Houhamdi L, Lepidi H, Drancourt M, Raoult D.** 2006. Experimental model to evaluate the human body louse as a vector of plague. *J Infect Dis* **194:**1589–1596. http://dx.doi.org/10.1086/508995.
2. **Bercovier M, Mollaret HH.** 1984. Yersinia, p 498–506. *In* Krieg NR, Holt JG (ed), *Bergey's Manual of Systematic Bacteriology,* **vol 1.** Williams & Wilkins, Baltimore, MD.
3. **Drancourt M, Houhamdi L, Raoult D.** 2006. *Yersinia pestis* as a telluric, human ectoparasite-borne organism. *Lancet Infect Dis* **6:**234–241. http://dx.doi.org/10.1016/S1473-3099(06)70438-8.
4. **Simond PL.** 1898. La propagation de la peste. *Ann Inst Pasteur (Paris)* **10:**625–687.
5. **Flatau G, Lemichez E, Gauthier M, Chardin P, Paris S, Fiorentini C, Boquet P.** 1997. Toxin-induced activation of the G protein p21 Rho by deamidation of glutamine. *Nature* **387:**729–733. http://dx.doi.org/10.1038/42743.
6. **Baltazard M, Aslani P.** 1952. [Biochemical characteristics of the strains of wild plague in Kurdistan]. *Ann Inst Pasteur (Paris)* **83:**241–247.
7. **Bitam I, Ayyadurai S, Kernif T, Chetta M, Boulaghman N, Raoult D, Drancourt M.** 2010. New rural focus of plague, Algeria. *Emerg Infect Dis* **16:**1639–1640. http://dx.doi.org/10.3201/eid1610.091854.
8. **Baltazard M.** 1964. The conservation of plague in inveterate foci. *J Hyg Epidemiol Microbiol Immunol* **8:**409–421.
9. **Baltazard M, Mofidi C.** 1950. Sur la peste inapparente des rongeurs sauvages. *C R Acad Sci* **231:**731–733.
10. **Baltazard M, Karimi Y, Eftekhari M, Chamsa M, Mollaret HH.** 1963. The interepizootic preservation of plague in an inveterate focus. Working hypotheses. *Bull Soc Pathol Exot Filiales* **56:**1230–1245.
11. **Blanc G, Baltazard M.** 1941. Recherches expérimentales sur la peste. L'infection du pou de l'homme *Pediculus corporis* de Geer. *C R Acad Sci* **213:**849–851.
12. **Mollaret HH.** 1963. Experimental preservation of plague in soil. *Bull Soc Pathol Exot Filiales* **56:**1168–1182.
13. **Begier EM, Asiki G, Anywaine Z, Yockey B, Schriefer ME, Aleti P, Ogden-Odoi A, Staples JE, Sexton C, Bearden SW, Kool JL.** 2006. Pneumonic plague cluster, Uganda, 2004. *Emerg Infect Dis* **12:**460–467. http://dx.doi.org/10.3201/eid1203.051051.
14. **Drancourt M, Aboudharam G, Signoli M, Dutour O, Raoult D.** 1998. Detection of 400-year-old *Yersinia pestis* DNA in human dental pulp: an approach to the diagnosis of ancient septicemia. *Proc Natl Acad Sci U S A* **95:**12637–12640. http://dx.doi.org/10.1073/pnas.95.21.12637.
15. **Aboudharam G, Drancourt M, Raoult D.** 2004. Culture of *C. burnetii* from the dental pulp of experimentally infected guinea pigs. *Microb Pathog* **36:**349–350. http://dx.doi.org/10.1016/j.micpath.2004.02.002.
16. **Aboudharam G, Lascola B, Raoult D, Drancourt M.** 2000. Detection of *Coxiella burnetii* DNA in dental pulp during experimental bacteremia. *Microb Pathog* **28:**249–254. http://dx.doi.org/10.1006/mpat.1999.0343.
17. **Raoult D, Aboudharam G, Crubézy E, Larrouy G, Ludes B, Drancourt M.** 2000. Molecular identification by "suicide PCR" of *Yersinia pestis* as the agent of medieval black death. *Proc Natl Acad Sci U S A* **97:**12800–12803. http://dx.doi.org/10.1073/pnas.220225197.
18. **Duncan CJ, Scott S.** 2005. What caused the Black Death? *Postgrad Med J* **81:**315–320. http://dx.doi.org/10.1136/pgmj.2004.024075.
19. **Scott S, Duncan CJ, Duncan SR.** 1996. The plague in Penrith, Cumbria, 1597/8: its causes, biology and consequences. *Ann Hum Biol* **23:**1–21. http://dx.doi.org/10.1080/03014469600004232.

20. **Gilbert MT, Cuccui J, White W, Lynnerup N, Titball RW, Cooper A, Prentice MB.** 2004. Absence of *Yersinia pestis*-specific DNA in human teeth from five European excavations of putative plague victims. *Microbiology* **150:**341–354. http://dx.doi.org/10.1099/mic.0.26594-0.

21. **Paterson R.** 2002. *Yersinia* seeks pardon for Black Death. *Lancet Infect Dis* **2:**323. http://dx.doi.org/10.1016/S1473-3099(02)00309-2.

22. **Raoult D, Drancourt M.** 2002. Cause of black death. *Lancet Infect Dis* **2:**459. http://dx.doi.org/10.1016/S1473-3099(02)00341-9.

23. **Drancourt M, Signoli M, Dang LV, Bizot B, Roux V, Tzortzis S, Raoult D.** 2007. *Yersinia pestis Orientalis* in remains of ancient plague patients. *Emerg Infect Dis* **13:**332–333. http://dx.doi.org/10.3201/eid1302.060197.

24. **Vergnaud G.** 2005. *Yersinia pestis* genotyping. *Emerg Infect Dis* **11:**1317–1318; author reply 1318–1319.

25. **Schuenemann VJ, Bos K, DeWitte S, Schmedes S, Jamieson J, Mittnik A, Forrest S, Coombes BK, Wood JW, Earn DJ, White W, Krause J, Poinar HN.** 2011. Targeted enrichment of ancient pathogens yielding the pPCP1 plasmid of *Yersinia pestis* from victims of the Black Death. *Proc Natl Acad Sci U S A* **108:**E746–E752. http://dx.doi.org/10.1073/pnas.1105107108.

26. **Tran TN, Signoli M, Fozzati L, Aboudharam G, Raoult D, Drancourt M.** 2011. High throughput, multiplexed pathogen detection authenticates plague waves in medieval Venice, Italy. *PLoS One* **6:**e16735. http://dx.doi.org/10.1371/journal.pone.0016735.

27. **Tran TN, Forestier CL, Drancourt M, Raoult D, Aboudharam G.** 2011. Brief communication: co-detection of *Bartonella quintana* and *Yersinia pestis* in an 11th-15th burial site in Bondy, France. *Am J Phys Anthropol* **145:**489–494. http://dx.doi.org/10.1002/ajpa.21510.

28. **Haensch S, Bianucci R, Signoli M, Rajerison M, Schultz M, Kacki S, Vermunt M, Weston DA, Hurst D, Achtman M, Carniel E, Bramanti B.** 2010. Distinct clones of *Yersinia pestis* caused the Black Death. *PLoS Pathog* **6:**e1001134. http://dx.doi.org/10.1371/journal.ppat.1001134.

29. **Wiechmann I, Harbeck M, Grupe G.** 2010. *Yersinia pestis* DNA sequences in late medieval skeletal finds, Bavaria. *Emerg Infect Dis* **16:**1806–1807. http://dx.doi.org/10.3201/eid1611.100598.

30. **Tran TN, Raoult D, Drancourt M.** 2011. *Yersinia pestis* DNA sequences in late medieval skeletal finds, Bavaria. *Emerg Infect Dis* 17:955–957; author reply 957. http://dx.doi.org/10.3201/eid1705.101777.

31. **Bos KI, Schuenemann VJ, Golding GB, Burbano HA, Waglechner N, Coombes BK, McPhee JB, DeWitte SN, Meyer M, Schmedes S, Wood J, Earn DJ, Herring DA, Bauer P, Poinar HN, Krause J.** 2011. A draft genome of *Yersinia pestis* from victims of the Black Death. *Nature* **478:**506–510. http://dx.doi.org/10.1038/nature10549.

32. **Gilbert MT.** 2014. *Yersinia pestis*: one pandemic, two pandemics, three pandemics, more? *Lancet Infect Dis* **14:**264–265. http://dx.doi.org/10.1016/S1473-3099(14)70002-7.

33. **Wagner DM, Klunk J, Harbeck M, Devault A, Waglechner N, Sahl JW, Enk J, Birdsell DN, Kuch M, Lumibao C, Poinar D, Pearson T, Fourment M, Golding B, Riehm JM, Earn DJ, Dewitte S, Rouillard JM, Grupe G, Wiechmann I, Bliska JB, Keim PS, Scholz HC, Holmes EC, Poinar H.** 2014. *Yersinia pestis* and the plague of Justinian 541-543 AD: a genomic analysis. *Lancet Infect Dis* **14:**319–326. http://dx.doi.org/10.1016/S1473-3099(13)70323-2.

34. **Molak M, Ho SY.** 2011. Evaluating the impact of post-mortem damage in ancient DNA: a theoretical approach. *J Mol Evol* **73:**244–255. http://dx.doi.org/10.1007/s00239-011-9474-z.

35. **Nguyen-Hieu T, Aboudharam G, Drancourt M.** 2012. Heat degradation of eukaryotic and bacterial DNA: an experimental model for paleomicrobiology. *BMC Res Notes* **5:**528. http://dx.doi.org/10.1186/1756-0500-5-528.

36. **Ayyadurai S, Houhamdi L, Lepidi H, Nappez C, Raoult D, Drancourt M.** 2008. Long-term persistence of virulent *Yersinia pestis* in soil. *Microbiology* **154:**2865–2871. http://dx.doi.org/10.1099/mic.0.2007/016154-0.

37. **Adler CJ, Dobney K, Weyrich LS, Kaidonis J, Walker AW, Haak W, Bradshaw CJ, Townsend G, Soltysiak A, Alt KW, Parkhill J, Cooper A.** 2013. Sequencing ancient calcified dental plaque shows changes in oral microbiota with dietary shifts of the Neolithic and Industrial revolutions. *Nat Genet* **45:**450–455, 455e1.

38. **Butler T.** 2009. Plague into the 21st century. *Clin Infect Dis* **49:**736–742. http://dx.doi.org/10.1086/604718.

39. **Raoult D, Roux V.** 1999. The body louse as a vector of reemerging human diseases. *Clin Infect Dis* **29:**888–911. http://dx.doi.org/10.1086/520454.

40. **Fournier PE, Minnick MF, Lepidi H, Salvo E, Raoult D.** 2001. Experimental model of human body louse infection using green fluorescent protein-expressing *Bartonella quintana*. *Infect*

Immun **69**:1876–1879. http://dx.doi.org/10.1128/IAI.69.3.1876-1879.2001.

41. **Houhamdi L, Raoult D.** 2005. Excretion of living *Borrelia recurrentis* in feces of infected human body lice. *J Infect Dis* **191**:1898–1906. http://dx.doi.org/10.1086/429920.

42. **Houhamdi L, Fournier PE, Fang R, Raoult D.** 2003. An experimental model of human body louse infection with *Rickettsia typhi. Ann N Y Acad Sci* **990**:617–627. http://dx.doi.org/10.1111/j.1749-6632.2003.tb07436.x.

43. **Houhamdi L, Fournier PE, Fang R, Lepidi H, Raoult D.** 2002. An experimental model of human body louse infection with *Rickettsia prowazekii. J Infect Dis* **186**:1639–1646. http://dx.doi.org/10.1086/345373.

44. **Fang R, Houhamdi L, Raoult D.** 2002. Detection of *Rickettsia prowazekii* in body lice and their feces by using monoclonal antibodies. *J Clin Microbiol* **40**:3358–3363. http://dx.doi.org/10.1128/JCM.40.9.3358-3363.2002.

45. **Houhamdi L, Raoult D.** 2008. Different genes govern *Yersinia pestis* pathogenicity in *Caenorhabditis elegans* and human lice. *Microb Pathog* **44**:435–437. http://dx.doi.org/10.1016/j.micpath.2007.11.007.

46. **Raoult D, Ndihokubwayo JB, Tissot-Dupont H, Roux V, Faugere B, Abegbinni R, Birtles RJ.** 1998. Outbreak of epidemic typhus associated with trench fever in Burundi. *Lancet* **352**:353–358. http://dx.doi.org/10.1016/S0140-6736(97)12433-3.

47. **Raoult D, Dutour O, Houhamdi L, Jankauskas R, Fournier PE, Ardagna Y, Drancourt M, Signoli M, La VD, Macia Y, Aboudharam G.** 2006. Evidence for louse-transmitted diseases in soldiers of Napoleon's Grand Army in Vilnius. *J Infect Dis* **193**:112–120. http://dx.doi.org/10.1086/498534.

48. **Piarroux R, Abedi AA, Shako JC, Kebela B, Karhemere S, Diatta G, Davoust B, Raoult D, Drancourt M.** 2013. Plague epidemics and lice, Democratic Republic of the Congo. *Emerg Infect Dis* **19**:505–506. http://dx.doi.org/10.3201/eid1903.121542.

49. **Anonymous.** 1997. A large outbreak of epidemic louse-borne typhus in Burundi. *Wkly Epidemiol Rec* **72**:152–153.

50. **Raoult D, Roux V, Ndihokubwayo JB, Bise G, Baudon D, Marte G, Birtles R.** 1997. Jail fever (epidemic typhus) outbreak in Burundi. *Emerg Infect Dis* **3**:357–360. http://dx.doi.org/10.3201/eid0303.970313.

51. **Anonymous.** 1994. Epidemic typhus risk in Rwandan refugee camps. *Wkly Epidemiol Rec* **69**:259.

Sources of Materials for Paleomicrobiology

5

GÉRARD ABOUDHARAM[1]

Paleomicrobiology aims to establish the diagnosis of ancient infectious diseases from human or environmental ancient samples, including animal specimens dating back thousands of years but with conservation differences depending on the nature of the molecules (DNA or proteins) or conservation environmental conditions (1). This science, born of the multiple disciplines of medical microbiology, anthropology, history, and related sciences such as archaeozoology, initially aimed to highlight ancient pathogens and more recently ancient microbiota, including functional data such as antibiotic resistance (2). The works of Ruffer are among the oldest works (3–5), especially on the discovery of a solution for rehydrating the mummified tissue.

The major objective of paleomicrobiology is, through access to ancient pathogens, to better analyze their evolution, possibly virulence, but their adaptation to their habitat and their vectors. The analysis of the data obtained allows us to appreciate the evolution of these ancient pathogens and confirm microbiology diagnoses suspected only on the basis of historical and anthropological data when mass graves are analyzed. The natural history of infectious diseases can be reconstructed. It is deduced from the evolutionary data obtained from phylogenetic reconstructions or

[1]Aix Marseille Université, URMITE, UMR CNRS 7278, IRD 198, Inserm 1095, Faculté de Médecine, Marseille, France.
Paleomicrobiology of Humans
Edited by Michel Drancourt and Didier Raoult
© 2016 American Society for Microbiology, Washington, DC
doi:10.1128/microbiolspec.PoH-0016-2015

by directly analyzing genetic sequences dating back thousands of years. All these data are derived from ancient samples whose sources can be very different in nature; exploration techniques vary depending on the nature of the source.

TISSUE SOURCES

The long-term persistence of DNA after the death of an organism was demonstrated in old samples of Egyptian mummies in 1985 (6). Despite chemical and enzymatic degradation, the possibility of sequencing bacterial, viral, archaeal, and parasite DNA molecules persists over time. Limits result from either chemical degradation by hydrolysis (causing a break in the DNA chain) or oxidation, which results in DNA damage and failure of the complementary nucleotides to bind them. Enzymatic degradation has another effect: enzymes digest DNA and the remaining molecules, causing autolysis (7, 8). Despite this chemical and enzymatic degradation, it is still possible to obtain molecular information from DNA sequences from the contributory old samples. Among the tissues that can be used for the detection of old nucleic acids and that are potential sources of DNA are preserved animals, bones, fossil teeth, scrolls, amber, and even cave paintings. They can be classified into two major categories of samples used in the search for ancient microorganisms.

Environmental samples include soil and polar ice (9, 10) and also the remains of plants, insects, and animals (11). These samples are particularly interesting because the animals may be sources of microorganisms that are pathogenic for humans (zoonoses), and the detection of pathogens in samples of ancient animals helps to reveal the epidemiology of infectious diseases in past centuries (12).

Human samples can be taken from frozen tissues, such as those of the man of Tyrol (10), or from mummified tissue (6, 13). Mummified tissues are of interest; the detection of DNA

of *Trypanosoma cruzi* in a mummy dated to about 600 years ago suggested that *T. cruzi* infected humans in the pre-Columbian period (14), and ancient mummified human remains dated from 2000 BC to about AD 1400 identified Chagas disease without recognizable pathological anatomy changes (15). Human samples can also be made of fixed and embedded tissues for pathological anatomy, from which one can obtain information, or of coprolites, which are a recent research direction in paleomicrobiology, but it is the skeletal remains that constitute the largest source of tissue. Fortuitously discovered most of the time, bodies may be buried in mass graves called disaster graves. In other cases, bodies are buried separately in single or multiple graves. Mass graves at first evoke acts of war, but the absence of injury or gross lesions usually suggests an epidemic infectious disease.

Bone remains are most abundant in this type of grave, but the use of bone is not without technical problems, especially when DNA is being extracted. We proposed the use of dental pulp contained in the pulp cavity of the tooth. It has advantages based on the morphology of the tooth and its vascularization. During bacteremic disease, microorganisms that reach the dental pulp may persist inside. Moreover, dental pulp is protected from environmental contamination by enamel and dentine. Finally, dental pulp DNA extraction techniques are simple. Recently, several paleomicrobiology studies involved dental calculus, which consists of deposits of calcified oral biofilm. Dental calculus contains pathogens associated with infection of the oral cavity and respiratory tract (Table 1).

ENVIRONMENTAL SAMPLES

Ancient microorganisms can be detected in environmental samples, such as soil and polar ice, and also samples collected from ancient plants, insects, and animals (11, 16).

TABLE 1 **Environmental and host sources for paleomicrobiology**

Source	Type of specimen	References
Environment		
• **Ice, permafrost**	Bacterial DNA (*Serratia, Enterobacter, Klebsiella,* *Yersinia, Acinetobacter* spp.)	9
	Fungal DNA	17
	Vegetal DNA	20
• **Soil**	Vegetal flora DNA	18
• **Ectoparasite**	Bacterial DNA (*Bartonella quintana,* *Rickettsia prowazekii*)	21
Host		
• **Frozen tissues**		
Intestinal	Vegetal DNA	10
• **Ectoparasite**		
Hair from mummy	Ectoparasite DNA	22
• **Blood**	Bacterial DNA (*Mycobacterium tuberculosis*)	65, 66
• **Colon**		
Human mummified tissues	*Schistosoma haematobium* antigen	13
Stomach tissue frozen	Bacterial DNA (*Helicobacter pylori*)	28
Animal intestine	Bacterial DNA	19
• **Skin**		
Mummified tissue	Human DNA	6
	DNA from *Trypanosoma cruzi*	12, 14, 15
	Bacterial DNA (*M. tuberculosis*)	64
• **Bone**	Genomic DNA (*Mycobacterium leprae*)	35
	Bacterial DNA (*M. tuberculosis*)	36
	Bacterial DNA (*Yersinia pestis*)	37
	Bacterial DNA (*Brucella* spp.)	38
	Aspartic acid	41
	Human DNA	42, 43, 44, 47, 48, 49, 51, 52
	Quagga DNA (type of zebra)	45
• **Coprolite**		
Human	DNA from *Ascaris*	11
	DNA from tick	23
	DNA from *Enterobius vermicularis*	24
	Human microbiome	29, 59
	Bacterial DNA (*Haemophilus parainfluenzae*)	32
	Viral, bacterial, archaeal metagenome	33
Animal	Viral RNA	19
	Remains of parasite in shark coprolite	30
	DNA from *Nothrotheriops shastensis*	31
• **Dental pulp**	Bacterial DNA (*B. quintana, R. prowazekii*)	21, 55, 56, 57, 67
	Bacterial DNA (*Y. pestis*)	36
	Bacterial DNA (*M. tuberculosis*)	41
	Aspartic acid	43, 47, 51, 52
	Human DNA	53, 54, 58
	Human microbiome	61
• **Dental calculus**	Human microbiome	59, 60, 62, 63

Frozen Environmental Materials

Frozen material can be regarded as exceptional for its rarity and its unique properties because freezing prevents the degradation of tissue and of bacterial and viral DNA. In ice samples estimated to be 2,000 years old and collected from the Blue Mountain Island in the Canadian arctic archipelago, an environment uninhabited by humans, 50 coliforms randomly chosen for identification and susceptibility were cultured (9). Members of the bacterial genera *Serratia*, *Enterobacter*, *Klebsiella*, and *Yersinia* were found; speciation, however, was not conclusive. The presence of plasmids and genotypic data provided evidence that potentially ancient organisms from ice can be grown from very distant water samples. In another study of fungal diversity, based on samples of permanently frozen Siberian soil and dated to 300,000 to 400,000 years ago, a wide range of microorganisms in the Siberian tundra was discovered (17). We may also mention the systematic campaigns of environmental levies, which are intended to establish the taxonomic diversity of arctic plant assemblages during the last 50,000 years (18). Fungal DNA sequencing identified previously undescribed *Ascomycetes*, *Basidiomycetes*, and *Zygomycetes* organisms found in the soil. These sequences identified fungi belonging to 12 different classes and 10 different orders. From caribou dung dating back 700 years and found frozen, two viruses were characterized (ancient caribou feces-associated virus, or aCFV) and a partial RNA viral genome (ancient Northwest Territories cripavirus, or aNCV) (19). The hypothesis was suggested that these viruses came from plant material eaten by caribou or flying insects, and their conservation can be attributed to the protection of viral capsids at very cold temperatures. Repeated experiments in two different and distant laboratories have attested to their authenticity. Willerslev et al. also showed that DNA and amino acids in buried bodies can be recovered from sections of deep ice cores, allowing the reconstruction of past flora and fauna (20).

Ectoparasite Samples and Animals

Ectoparasites are important sources of information regarding living conditions and the transmission of infectious diseases. In the study conducted on the remains of the soldiers of Napoleon's army found in Lithuania in Vilnius, lice fragments were perfectly identified morphologically and by molecular biology (21). DNA from *Bartonella quintana* (the bacterial agent of trench fever) and *Rickettsia prowazekii* (the bacterial agent of epidemic typhus) was amplified by PCR and sequenced (21). The presence of these microorganisms, transmitted by the body louse, attests to the poor living conditions and overcrowding of the soldiers of the Grand Army, and in addition provides indisputable evidence of a hypothesis formulated by historians that many of the Grand Army soldiers died of typhus. Another work identified head lice nits on hairs of a 4,000-year-old Camarón Chilean mummy. Molecular analysis identified clade B head lice, which were therefore present in America before the arrival of European settlers (22).

The study of ectoparasites may give some information on historical vector-borne epidemics. More surprisingly, the identification of ectoparasites in a coprolite of human origin indicates the traditional lifestyles of the Anasazi: food habits indicating that ectoparasites were eaten by these ancient people and could have been a source of disease (23). Iniguez et al. used a molecular diagnostic and paleoparasitological approach and identified *Enterobius vermicularis* in coprolites from archaeological sites in Chile and the United States (24).

HUMAN SAMPLES

Frozen Human Tissues

Frozen human tissues constitutes an exceptional source for analysis. At very low temperatures, all biological activity, including biological reactions that would cause cell death, is suspended. However, the discovery of these sources is accidental, and they are relatively rare; often cited is the Ice Man found in Tyrol and named Ötzi. Data indicate that he was living in Central Europe about 5,000 years ago, at a time of transition between the Neolithic and the Copper Age. His body, dehydrated by freezing, was removed from an Alpine glacier in 1991, near the border between Austria and Italy and was thawed in 2000. Many studies have yielded a wealth of information on the development of demographic models (25) or on the body's origin and relationship to modern European populations (26) and the oral dental pathologies of that time (27).

Furthermore, bacteria can be found inside frozen bodies. For example, *Helicobacter pylori* DNA was sequenced from the stomach tissue of an individual found in the Samuel Glacier in Tatshenshini-Alsek Park, British Columbia, Canada (28). Strain *H. Tatshenshini pylori* vacA has a potential hybrid M2A / M1D. The vacA allele is more commonly found in Western strains, suggesting that European strains were present in Canada's northwest at the time when this individual lived (28).

Coprolites

Coprolites are a source for exploration of the human microbiome and some pathogens. Coprolites consist of mineralized, fossilized excrement. They are a constituent of sediment produced by the action of digestion of living beings. Studying coprolites allows us to estimate and evaluate changes of the human microbiome (29). For example, vertebrate coprolites dated to the Mesozoic Era and Paleozoic Era (66/541 mA) were shown to contain parasite remains characteristic of modern tapeworms. This is the earliest evidence of the presence of intestinal parasites in vertebrates (30). Also, a coprolite from the Pleistocene found in the Gypsum Cave in Nevada helped replenish the supply of a sloth that has disappeared today: *Nothrotheriops shastensis* (31). DNA from *Haemophilus parainfluenzae*

was identified in a stool sample collected directly from the intestines of a pre-Columbian mummy in Cuzco (Peru) and was estimated by radiocarbon dating 980 to 1170 AD, suggesting that this pathogen was present in pre-contact Native Americans (32). More recently, a 14th century medieval coprolite recovered from a site in Namur, Belgium, yielded a DNA viral metagenome (33). Its comparison with a modern human feces virome enhanced our understanding of past viral diversity and distribution (33). Coprolites are therefore a focus for the study and development of the human microbiome and in addition provide a very interesting way in which to study parasitic infections in humans in relation to ancient ecology, habitats, and diets.

Bone Tissues

Bone tissues are a material of choice because of their abundance. Bones have been used for the detection of the leprosy agent *Mycobacterium leprae* (34, 35), *Mycobacterium tuberculosis* (36), *Yersinia pestis* (37), and *Brucella* spp. (38). However, bone tissues may contain small amounts of DNA as a result of natural washing out. One idea that researchers propose is amino acid racemization. This technique provided the basis for a method of dating fossils (39), but more recently it has demonstrated a relationship between the rate of racemization of aspartic acid and the conservation of DNA (40, 41). Spectroscopy analysis also allows us to assess the degree of ancient tissue preservation in bone; it represents an effective appreciation element and allows us to choose one or another bone fragment according to the amount of DNA that is potentially available (42). Is joined the exceptional persistence of DNA in the time to the adsorption of DNA on the apatite crystals that comprise the mineral portion of bone but also to environmental factors such as temperature and/or a dry environment. The authors deduced that preservation of the apatite phase can dramatically affect the recovery of DNA (43).

Despite the widespread use of bones in studies of ancient DNA, relatively little concrete information exists regarding how DNA degrades over mineralized collagen or if it survives in architecture matter. At the structural level: physical exclusion microbes and other external contaminants can be an important characteristic; at ultrastructural level the DNA adsorption to hydroxyapatite, the relative contribution of each factor is unclear. For example, there is considerable variation in the quality of the DNA extracted from bones and teeth. This is partly due to various environmental factors such as temperature, proximity to free water or oxygen, pH, salt content, and radiation exposure, which increases the decay rate of DNA (44).

The extraction of DNA from ancient bone is not an easy task; when skeletons are in graves in the ground, their DNA is generally rare and poorly maintained. This poor preservation of DNA is partly related to sediment. The conditions are more favorable when environmental conditions are better, such as when skeletons are buried in vaults. The risk for contamination is significantly lower. It is also necessary to note the presence of PCR-inhibiting enzymes such as humic acids, tannins, and fulvic acids (45, 46). Another problem that arises during studies of skeletal remains is that DNA must be separated from the mineral fraction; therefore, decalcifying buffers such as ethylenediamine tetraacetic (EDTA) must be used, which may then decrease the performance of PCR.

Various studies have proposed technical solutions to overcome this drawback (47–49) and increase the efficiency of extraction. To overcome the problems associated with microbial DNA extraction from bony remains, we have developed in the laboratory a novel technique of dental pulp recovery from fossil teeth and directed our pathogen detection work this original art by this original method from the pulp.

Teeth

After burial, teeth may persist longer than bones and are therefore a particularly interesting tissue source. The tooth structure consists of heavily mineralized tissues (70% to 75% of the dry weight of mineralized teeth compared with 64% for bone) (Fig. 1), so that teeth deteriorate over time more slowly than bone (50).

Researchers have in the past used teeth to search for ancient DNA. The techniques used were interesting, but the authors did not take full advantage of these bodies because they had pulverized the entire teeth for their studies (51–53). In this way, the difficulty of extracting the DNA is identical to that encountered when DNA is extracted from bone

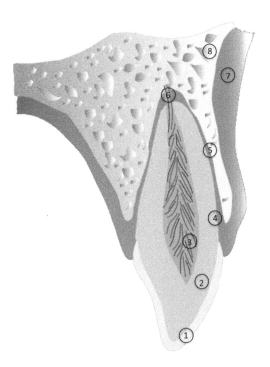

FIGURE 1 **Cross-section of a tooth. 1, enamel; 2, dentin; 3, dental pulp; 4, cement; 5, periodontal fiber; 6, apical foramen; 7, gingival mucosa; 8, alveolar bone.**

because it is necessary to remove the mineral part. However, Ginther et al. managed to amplify mitochondrial DNA (mtDNA) from teeth and concluded that teeth could be an excellent source of high-molecular-weight mtDNA (54). This could be useful to extend the period within which decomposed human remains could be genetically identified. We then proposed to use dental pulp instead of the entire tooth for paleomicrobiology (55, 56).

The arguments in favor of the use of dental pulp stem from the fact that pulp is a vascular connective tissue mass that occupies the central pulp cavity of the tooth. Like all connective tissue, pulp is formed fundamentally of cellular elements and fibers. This connective tissue is as highly vascularized as the brain, indicating the probability that bacteremic microorganisms were able to persist in the dental pulp remains and residual blood after the death of the individual is important. Also, potential dental pulp remains are pulp remnants are extremely well preserved from natural contamination and wash sediment. An important argument for using pulp remains for paleomicrobiology is that DNA extraction techniques are much simpler. Indeed, it is not necessary to go through the steps of demineralization to extract DNA from pulp remnants as it is with bone, which is organic material. We recently showed that with the use of our protocol, dental pulp is not prone to in-laboratory contamination (M. Drancourt, G. Aboudharam, unpublished data).

If the dental pulp recovery techniques vary according to the authors (55, 57, 58), opening of the pulp cavity along the longest axis of the tooth is the most reliable. Solidarity jaw teeth are more likely to have been preserved from external contamination and more wisdom tooth.

Dental Calculus

Dental calculus forms a calcified oral biofilm deposited on the teeth after the complex process of mineralization. A combination of endogenous factors, such as oral hygiene, the amount of saliva, and its pH and buffing capacity, and exogenous factors, such as exposure to smoking or radiation and drugs that decrease salivary flow, results in a greater or lesser formation of calculus. Oral microorganisms are trapped and kept within this mineralized matrix. It contains the pathogens associated with the oral cavity and respiratory tract. Therefore, dental calculus is a particularly interesting source for the study of the ancient oral microbiome (59, 60) and pathogens associated with oral diseases (61). From the dental calculus of four adult human skeletons from the site of Dalheim, Germany, and dated to 950-1200 BC, the author provided proof of periodontal disease in these individuals and applying shotgun DNA sequencing, established the taxonomic and phylogenetic characterization of the ancient calculus (62). Major dietary changes occurred during human history when a diet based on agriculture changed to a carbohydrate-rich diet with industrialization around 1850. This has led to a significant change in the oral flora and has been studied and demonstrated in dental calculus from 34 European skeletons. The study showed that the oral microbiome composition remained stable from the Neolithic Age to the Middle Ages and evolved with industrialization to a configuration associated with the disease (63). Obviously, the study of ancient calculus as a new potential source of ancient microbial DNA is promising for understanding the evolution of the oral microbiome and associated pathologies.

HANDLING AND PRESERVING ANCIENT SPECIMENS FOR PALEOMICROBIOLOGY

Precautions during Excavations:

Generally, archaeological features and particularly human remains (teeth, jaws, or skulls) that are to be the subject of subsequent molecular analyses are collected to avoid contamination with contemporary human and microbial DNA. These precautions are based mainly on the use of gloves, masks, and caps. When the elements are removed, administrative information about the samples is first collected: total number, date of discovery, full name of the archaeologist or author of the levy, and additional comments (if necessary). All samples are packaged in containers of sufficient strength (individual bags in general) so that objects cannot be lost, damaged, or mixed with other objects, particularly during transport to the laboratory.

Location of Individuals in the Grave

For purposes of analysis, individuals are selected if possible in areas of the best-preserved body after the location of mass graves and topographic surveys (work devolved to the anthropologists managing the excavation). Priority skulls and jawbones with the largest number of teeth are chosen. If the teeth are separated from arcades, they are associated with an individual or considered anonymous.

Choice of Teeth

All operations are done in order to prevent extrinsic contamination and also to protect the operators from possible contamination by sampling. To date, no contamination of an operator on a mass grave has been described; however, care must be taken to avoid potential contaminants. Indeed, in an attempt to recover *M. tuberculosis* from mummified tissue about 300 years old, although no culture was observed, *M. tuberculosis* was detected by PCR, suggesting that the survival of *M. tuberculosis* is less than a few hundred years (64). Infection with *M. tuberculosis* of a thanatopracteur in 2007, however, was reported after manipulation of the body of a patient who died in October 2005; the embalmer was contaminated by the deceased because he was before negative for tuberculin

tests. The results confirmed the genotype transmission (65). In another case of contamination of a thanatopracteur, the embalmer usually used gloves and a mask (66). However, these cases represent contamination during the manipulation of people whose death was recent. To our knowledge, no contamination has been reported to date originating from human remains dated from several hundred years.

The inventiveness of researchers trying to grow bacteria from ancient remains in order to obtain information to understand the evolution of microbes is impressive. Culture attempts include those to culture *Y. pestis* from ancient dental pulp (Drancourt M, Aboudharam G, Raoult D. URMITE Marseille, unpublished data). These attempts have been unsuccessful to date. However, all samples are processed in the laboratory as if the pathogen were "sleeping."

Treatment of Samples

When samples are received or delivered to the laboratory, the implementation procedure includes recording in the database of samples entering the laboratory, for purposes of traceability. The samples are then photographed, X-rayed, and stored.

- Samples are stored at room temperature or, more precisely, at temperature conditions as similar as possible to the conditions in which the remains of the individuals were found. The storage of solid samples (mineralized bone and tooth) is at room temperature, and soft tissue is stored at –20°C.
- All samples are photographed. The photographs are a medium of information and research for anthropologists. These pictures can later be used for further studies. Furthermore, pictures taken from all sides in the case of teeth enable measurements for statistical purposes.
- Radiography makes it possible to assess the importance of the pulp cavity. The

most suitable teeth for these studies are the single-rooted teeth with a large pulp cavity (incisors, canines, and premolars) of young adults or adolescents that have an apical end that is relatively closed. If it is open, the dental pulp cavity may have been washed by the sediment and dental pulp remains degraded.

ETHICS AND LEGAL FRAMEWORK

The studies must first be approved by the ethics committee overseeing the research. In general, information on human remains is provided by anthropologists and samples treated anonymously. Under French law, no authorization is required for scientific investigations of old graves. According to French law number 2001-44 of 17 January 2001 on preventive archeology (http://legifrance.gouv.fr/), ancient human remains are considered archaeological remains, like other remains (e.g., pottery, coins). Anthropologists and archaeologists, because of their position, are free to take samples and analyze them for their scientific research by any technique they need. Anecdotally may be mentioned the only work done to date on a clearly identified body. This was the discovery of *Y. pestis* in the teeth of a body found in 1986 in a lead sarcophagus. The interred was an English nobleman, Thomas Craven, a Protestant and member of the Paris Reformed Church, who died in 1636 at the age of 18 years (67).

CITATION

Aboudharam G. 2016. Sources of Materials for Paleomicrobiology, Microbiol Spectrum 4(4):PoH-0016-2015.

REFERENCES

1. **Anastasiou E, Mitchell PD.** 2013. Palaeopathology and genes: investigating the genetics of infectious diseases in excavated human skeletal remains and mummies from past populations. *Gene* **528:**33–40.

2. **Olaitan A, Rolain JM.** 2016. Ancient resistome. *In* Raoult D, Drancourt M (ed), *Paleomicrobiology of Humans.* ASM Press, Washington, DC. (In press)

3. **Ruffer MA.** 1913. On pathological lesions found in Coptic bodies (400-500 AD). *J Pathol Bacteriol* **18:**149–162.

4. **Ruffer MA.** 1921. *Studies in the Palaeopathology of Egypt.* University of Chicago Press, Chicago, IL.

5. **Sandison AT.** 1967. Sir Marc Armand Ruffer (1859-1917) pioneer of palaeopathology. *Med Hist* **11:**150–156.

6. **Pääbo S.** 1985. Molecular cloning of ancient Egyptian mummy DNA. *Nature* **314:**644–645.

7. **Nguyen-Hieu T, Aboudharam G, Drancourt M.** 2012. Heat degradation of eukaryotic and bacterial DNA: an experimental model for paleomicrobiology. *BMC Res Notes* **5:**528.

8. **Lindahl T.** 1993. Instability and decay of the primary structure of DNA. *Nature* **362:**709–715.

9. **Dancer SJ, Shears P, Platt DJ.** 1997. Isolation and characterization of coliforms from glacial ice and water in Canada's High Arctic. *J Appl Microbiol* **82:**597–609.

10. **Rollo F, Ubaldi M, Ermini L, Marota I.** 2002. Otzi's last meals: DNA analysis of the intestinal content of the Neolithic glacier mummy from the Alps. *Proc Natl Acad Sci U S A* **99:** 12594–12599.

11. **Loreille O, Roumat E, Verneau O, Bouchet F, Hänni C.** 2001. Ancient DNA from Ascaris: extraction amplification and sequences from eggs collected in coprolites. *Int J Parasitol* **31:** 1101–1106.

12. **Ferreira LF, Britto C, Cardoso MA, Fernandes O, Reinhard K, Araújo A.** 2000. Paleoparasitology of Chagas disease revealed by infected tissues from Chilean mummies. *Acta Trop* **75:**79–84.

13. **Deelder AM, Miller RL, de Jonge N, Krijger FW.** 1990. Detection of schistosome antigen in mummies. *Lancet* **335:**724–725.

14. **Fernandes A, Iñiguez AM, Lima VS, Souza SM, Ferreira LF, Vicente AC, Jansen AM.** 2008. Pre-Columbian Chagas disease in Brazil: *Trypanosoma cruzi* I in the archaeological remains of a human in Peruaçu Valley, Minas Gerais, Brazil. *Mem Inst Oswaldo Cruz* **103:** 514–516.

15. **Guhl F, Jaramillo C, Vallejo GA, Yockteng R, Cárdenas-Arroyo F, Fornaciari G, Arriaza B, Aufderheide AC.** 1999. Isolation of *Trypanosoma cruzi* DNA in 4,000-year-old

mummified human tissue from northern Chile. *Am J Phys Anthropol* **108**:401–407.

16. Rhodes AN, Urbance JW, Youga H, Corlew-Newman H, Reddy CA, Klug MJ, Tiedje JM, Fisher DC. 1998. Identification of bacterial isolates obtained from intestinal contents associated with 12,000-year-old mastodon remains. *Appl Environ Microbiol* **64**:651–658.

17. Lydolph MC, Jacobsen J, Arctander P, Gilbert MT, Gilichinsky DA, Hansen AJ, Willerslev E, Lange L. 2005. Beringian paleoecology inferred from permafrost-preserved fungal DNA. *Appl Environ Microbiol* **71**:1012–1017.

18. Willerslev E, Davison J, Moora M, Zobel M, Coissac E, Edwards ME, Lorenzen ED, Vestergård M, Gussarova G, Haile J, Craine J, Gielly L, Boessenkool S, Epp LS, Pearman PB, Cheddadi R, Murray D, Bråthen KA, Yoccoz N, Binney H, Cruaud C, Wincker P, Goslar T, Alsos IG, Bellemain E, Brysting AK, Elven R, Sønstebø JH, Murton J, Sher A, Rasmussen M, Rønn R, Mourier T, Cooper A, Austin J, Möller P, Froese D, Zazula G, Pompanon F, Rioux D, Niderkorn V, Tikhonov A, Savvinov G, Roberts RG, MacPhee RD, Gilbert MT, Kjær KH, Orlando L, Brochmann C, Taberlet P. 2014. Fifty thousand years of Arctic vegetation and megafaunal diet. *Nature* **506**:47–51.

19. Ng TF, Chen LF, Zhou Y, Shapiro B, Stiller M, Heintzman PD, Varsani A, Kondov NO, Wong W, Deng X, Andrews TD, Moorman BJ, Meulendyk T, MacKay G, Gilbertson RL, Delwart E. 2014. Preservation of viral genomes in 700-y-old caribou feces from a subarctic ice patch. *Proc Natl Acad Sci U S A* **111**:16842–16847.

20. Willerslev E, Cappellini E, Boomsma W, Nielsen R, Hebsgaard MB, Brand TB, Hofreiter M, Bunce M, Poinar HN, Dahl-Jensen D, Johnsen S, Steffensen JP, Bennike O, Schwenninger JL, Nathan R, Armitage S, de Hoog CJ, Alfimov V, Christl M, Beer J, Muscheler R, Barker J, Sharp M, Penkman KE, Haile J, Taberlet P, Gilbert MT, Casoli A, Campani E, Collins MJ. 2007. Ancient biomolecules from deep ice cores reveal a forested southern Greenland. *Science* **317**:111–114.

21. Raoult D, Dutour O, Houhamdi L, Jankauskas R, Fournier PE, Ardagna Y, Drancourt M, Signoli M, La VD, Macia Y, Aboudharam G. 2006. Evidence for louse-transmitted diseases in soldiers of Napoleon's Grand Army in Vilnius. *J Infect Dis* **193**:112–120.

22. Boutellis A, Drali R, Rivera MA, Mumcuoglu KY, Raoult D. 2013. Evidence of sympatry of clade A and clade B head lice in a pre-Columbian Chilean mummy from Camarones. *PLoS One* **8**:e76818.

23. Johnson KL, Reinhardt KJ, Sianto L, Araujo A, Gardner SL, Janovy J Jr. 2008. A tick from a prehistoric Arizona coprolite. *J Parasitol* **94**:296–298.

24. Iniguez AM, Reinhard KJ, Araujo A, Ferreira LF, Vicente AC. 2003. *Enterobius vermicularis*: ancient DNA from North and South American human coprolites. *Mem Inst Oswaldo Cruz* **98**(Suppl 1):67–69.

25. Sams AJ, Hawks J, Keinan A. 2015. The utility of ancient human DNA for improving allele age estimates, with implications for demographic models and tests of natural selection. *J Hum Evol* **79**:64–72.

26. Sikora M, Carpenter ML, Moreno-Estrada A, Henn BM, Underhill PA, Sánchez-Quinto F, Zara I, Pitzalis M, Sidore C, Busonero F, Maschio A, Angius A, Jones C, Mendoza-Revilla J, Nekhrizov G, Dimitrova D, Theodossiev N, Harkins TT, Keller A, Maixner F, Zink A, Abecasis G, Sanna S, Cucca F, Bustamante CD. 2014. Population genomic analysis of ancient and modern genomes yields new insights into the genetic ancestry of the Tyrolean Iceman and the genetic structure of Europe. *PLoS Genet* **10**: e1004353. doi:10.1371/journal.pone.0076818.

27. Seiler R, Spielman AI, Zink A, Rühli F. 2013. Oral pathologies of the Neolithic Iceman, c.3,300 BC. *Eur J Oral Sci* **121**:137–141.

28. Swanston T, Haakensen M, Deneer H, Walker EG. 2011. The characterization of *Helicobacter pylori* DNA associated with ancient human remains recovered from a Canadian glacier. *PLoS One* **6**:e16864. doi:10.1371/journal.pone.0016864.

29. Tito RY, Knights D, Metcalf J, Obregon-Tito AJ, Cleeland L, Najar F, Roe B, Reinhard K, Sobolik K, Belknap S, Foster M, Spicer P, Knight R, Lewis CM Jr. 2012. Insights from characterizing extinct human gut microbiomes. *PLoS One* **7**:e51146. doi:10.1371/journal.pone.0051146.

30. Dentzien-Dias PC, Poinar G Jr, de Figueiredo AE, Pacheco AC, Horn BL, Schultz CL. 2013. Tapeworm eggs in a 270 million-year-old shark coprolite. *PLoS One* **8**:e55007. doi:10.1371/journal.pone.0055007.

31. Poinar HN, Hofreiter M, Spaulding WG, Martin PS, Stankiewicz BA, Bland H, Evershed RP, Possnert G, Pääbo S. 1998. Molecular coproscopy: dung and diet of the

extinct ground sloth *Nothrotheriops shastensis*. *Science* **281**:402–406.

32. **Luciani S, Fornaciari G, Rickards O, Labarga CM, Rollo F.** 2006. Molecular characterization of a pre-Columbian mummy and in situ coprolite. *Am J Phys Anthropol* **129**:620–629.

33. **Appelt S, Fancello L, Le Bailly M, Raoult D, Drancourt M, Desnues C.** 2014. Viruses in a 14th-century coprolite. *Appl Environ Microbiol* **80**:2648–2655.

34. **Mendum TA, Schuenemann VJ, Roffey S, Taylor GM, Wu H, Singh P, Tucker K, Hinds J, Cole ST, Kierzek AM, Nieselt K, Krause J, Stewart GR.** 2014. *Mycobacterium leprae* genomes from a British medieval leprosy hospital: towards understanding an ancient epidemic. *BMC Genomics* **15**:270.

35. **Schuenemann VJ, Singh P, Mendum TA, Krause-Kyora B, Jäger G, Bos KI, Herbig A, Economou C, Benjak A, Busso P, Nebel A, Boldsen JL, Kjellström A, Wu H, Stewart GR, Taylor GM, Bauer P, Lee OY, Wu HH, Minnikin DE, Besra GS, Tucker K, Roffey S, Sow SO, Cole ST, Nieselt K, Krause J.** 2013. Genome-wide comparison of medieval and modern *Mycobacterium leprae*. *Science* **341**:179–183.

36. **Muller R, Roberts CA, Brown TA.** 2014. Genotyping of ancient *Mycobacterium tuberculosis* strains reveals historic genetic diversity. *Proc Biol Sci* **281**:20133326.

37. **Seifert L, Harbeck M, Thomas A, Hoke N, Zöller L, Wiechmann I, Grupe G, Scholz HC, Riehm JM.** 2013. Strategy for sensitive and specific detection of *Yersinia pestis* in skeletons of the black death pandemic. *PLoS One* **8**:e75742. doi:10.1371/journal.pone.0075742.

38. **Mutolo MJ, Jenny LL, Buszek AR, Fenton TW, Foran DR.** 2012. Osteological and molecular identification of Brucellosis in ancient Butrint, Albania. *Am J Phys Anthropol* **147**:254–263.

39. **Bada JL.** 1972. Kinetics of racemization of amino acids as a function of pH. *J Am Chem Soc* **94**:1371–1373.

40. **Poinar HN, Hoss M, Bada JL, Paabo S.** 1996. Amino acid racemization and the preservation of ancient DNA. *Science* **272**:864–866.

41. **Collins MJ, Penkman KE, Rohland N, Shapiro B, Dobberstein RC, Ritz-Timme S, Hofreiter M.** 2009. Is amino acid racemization a useful tool for screening for ancient DNA in bone? *Proc Biol Sci* **276**:2971–2977.

42. **Scorrano G, Valentini F, Martinez-Labarga C, Rolfo MF, Fiammenghi A, Lo Vetro D, Martini F, Casoli A, Ferraris G, Palleschi G,**

43. **Palleschi A, Rickards O.** 2015. Methodological strategies to assess the degree of bone preservation for ancient DNA studies. *Ann Hum Biol* **42**:10–19.

44. **Grunenwald A, Keyser C, Sautereau AM, Crubezy E, Ludes B, Drouet C.** 2014. Novel contribution on the diagenetic physicochemical features of bone and teeth minerals, as substrates for ancient DNA typing. *Anal Bioanal Chem* **406**:4691–4704.

45. **Campos PF, Craig OE, Turner-Walker G, Peacock E, Willerslev E, Gilbert MT.** 2012. DNA in ancient bone - where is it located and how should we extract it? *Ann Anat* **194**:7–16.

46. **Pääbo S, Poinar H, Serre D, Jaenicke-Despres V, Hebler J, Rohland N, Kuch M, Krause J, Vigilant L, Hofreiter M.** 2004. Genetic analyses from ancient DNA. *Annu Rev Genet* **38**:645–679.

47. **O'Rourke DH, Hayes MG, Carlyle SW.** 2000. Spatial and temporal stability of mtDNA haplogroup frequencies in native North America. *Hum Biol* **72**:15–34.

48. **Rohland N, Hofreiter M.** 2007. Ancient DNA extraction from bones and teeth. *Nat Protoc* **2**:1756–1762.

49. **Rohland N, Hofreiter M.** 2007. Comparison and optimization of ancient DNA extraction. *Biotechniques* **42**:343–352.

50. **Dukes MJ, Williams AL, Massey CM, Wojtkiewicz PW.** 2012. Technical note: Bone DNA extraction and purification using silica-coated paramagnetic beads. *Am J Phys Anthropol* **148**:473–482.

51. **Kirsanow K, Burger J.** 2012. Ancient human DNA. *Ann Anat* **194**:121–132.

52. **Kurosaki K, Matsushita T, Ueda S.** 1993. Individual DNA identification from ancient human remains. *Am J Hum Genet* **53**:638–643.

53. **Hanni C, Laudet V, Sakka M, Begue A, Stehelin D.** 1990. [Amplification of mitochondrial DNA fragments from ancient human teeth and bones]. *C R Acad Sci III* **310**:365–370.

54. **Hummel S, Schultes T, Bramanti B, Herrmann B.** 1999. Ancient DNA profiling by megaplex amplications. *Electrophoresis* **20**:1717–1721.

55. **Ginther C, Issel-Tarver L, King MC.** 1992. Identifying individuals by sequencing mitochondrial DNA from teeth. *Nat Genet* **2**:135–138.

56. **Drancourt M, Aboudharam G, Signoli M, Dutour O, Raoult D.** 1998. Detection of 400-year-old *Yersinia pestis* DNA in human dental pulp: an approach to the diagnosis of ancient septicemia. *Proc Natl Acad Sci U S A* **95**:12637–12640.

56. **Raoult D, Aboudharam G, Crubézy E, Larrouy G, Ludes B, Drancourt M.** 2000. Molecular identification by "suicide PCR" of *Yersinia pestis* as the agent of medieval black death. *Proc Natl Acad Sci U S A* **97**:12800–12803.

57. **Tran-Hung L, Tran-Thi N, Aboudharam G, Raoult D, Drancourt M.** 2007. A new method to extract dental pulp DNA: application to universal detection of bacteria. *PLoS One* **2**: e1062. doi:10.1371/journal.pone.0001062.

58. **Hervella M, Iniguez MG, Izagirre N, Anta A, De-la-Rua C.** 2015. Nondestructive methods for recovery of biological material from human teeth for DNA extraction. *J Forensic Sci* **60**:136–141.

59. **Warinner C, Speller C, Collins MJ, Lewis CM Jr.** 2015. Ancient human microbiomes. *J Hum Evol* **79c**:125–136.

60. **Warinner C, Speller C, Collins MJ.** 2015. A new era in palaeomicrobiology: prospects for ancient dental calculus as a long-term record of the human oral microbiome. *Philos Trans R Soc Lond B Biol Sci* **370**:20130376.

61. **Curry A.** 2013. Ancient DNA. Fossilized teeth offer mouthful on ancient microbiome. *Science* **342**:1303.

62. **Warinner C, Rodrigues JF, Vyas R, Trachsel C, Shved N, Grossmann J, Radini A, Hancock Y, Tito RY, Fiddyment S, Speller C, Hendy J, Charlton S, Luder HU, Salazar-García DC, Eppler E, Seiler R, Hansen LH, Castruita JA, Barkow-Oesterreicher S, Teoh KY, Kelstrup CD, Olsen JV, Nanni P, Kawai T, Willerslev E, von Mering C, Lewis CM Jr, Collins MJ, Gilbert MT, Rühli F, Cappellini E.** 2014. Pathogens and host immunity in the ancient human oral cavity. *Nat Genet* **46**:336–344.

63. **Adler CJ, Dobney K, Weyrich LS, Kaidonis J, Walker AW, Haak W, Bradshaw CJ, Townsend G, Sołtysiak A, Alt KW, Parkhill J, Cooper A.** 2013. Sequencing ancient calcified dental plaque shows changes in oral microbiota with dietary shifts of the Neolithic and Industrial revolutions. *Nat Genet* **45**:450–455, 455e451. doi:10.1038/ng.2536.

64. **Lemma E, Zimhony O, Greenblatt CL, Koltunov V, Zylber MI, Vernon K, Spigelman M.** 2008. Attempts to revive *Mycobacterium tuberculosis* from 300-year-old human mummies. *FEMS Microbiol Lett* **283**:54–61.

65. **Anderson J, Meissner J, Ahuja S, Shashkina E, O'Flaherty T, Proops DC.** 2015. Confirming *Mycobacterium tuberculosis* transmission from a cadaver to an embalmer using molecular epidemiology. *Am J Infect Control* **43**:543–545.

66. **Sterling TR, Pope DS, Bishai WR, Harrington S, Gershon RR, Chaisson RE.** 2000. Transmission of *Mycobacterium tuberculosis* from a cadaver to an embalmer. *N Engl J Med* **342**:246–248.

67. **Hadjouis D, La Vu D, Aboudharam G, Drancourt M, Andrieux P.** 2008. Thomas Craven, noble anglais mort de la peste en 1636 à Saint-Maurice (Val-de-Marne, France). Identification et détermination de la cause de la mort par l'ADN. *Biometrie Humaine et Anthropologie* **26**:69–76.

Paleomicrobiology Data: Authentification and Interpretation

6

MICHEL DRANCOURT[1]

Paleomicrobiology is a part of general microbiology that aims to describe microbial flora (including bacteria, archaea, viruses, parasites, and microscopic fungi) in older specimens; it includes the retrospective diagnosis of infectious and tropical diseases (1, 2). There is no precise numerical definition of an old specimen; we consider as paleomicrobiological specimens any specimen older than 100 years, while the study of more recent samples may be called meso-microbiology. Also, we will consider in this chapter only animal and human specimens, excluding inanimate environmental specimens. The initial works of paleopathology, which used molecular biology techniques based on PCR, highlighted the possibility of the contamination of old samples by contemporary human DNA, casting doubt on the validity of PCR-based work for the detection of human and microbial ancient DNA (aDNA). These doubts, arising in the field of human aDNA, were then relayed to the field of paleomicrobiology by colleagues not practicing microbiology and without microbiological knowledge, who simply derived doubts from their observations of old human DNA. In fact, microbial contamination of an ancient specimen can take place *in situ* from the surrounding flora, which can be either ancient flora contemporary with the sample or flora from operators at the time of excavation

[1]Aix-Marseille Université, URMITE (Unité de Recherche sur les Maladies Infectieuses et Tropicales Emergentes), UMR CNRS 7278, IRD 198, Inserm 1095, Institut Hospitalier Universitaire Méditerranée Infection, Marseille, France.

Paleomicrobiology of Humans
Edited by Michel Drancourt and Didier Raoult
© 2016 American Society for Microbiology, Washington, DC
doi:10.1128/microbiolspec.PoH-0017-2015

and storage of the old material or during laboratory manipulations, which is modern contamination. Contamination can occur with whole organisms or biomolecules of interest, such as nucleic acids or proteins or mycolic acids (Fig. 1). Indeed, several methods are now used for discovering microbes in ancient specimens beyond the now-conventional aDNA analyses (2).

We herein review the possible sources of contamination of samples by ancient contemporary microorganisms, the methods for detecting such potential contamination, and the methods for preventing it.

TAPHONOMICAL CONTAMINATION OF ANCIENT SPECIMENS

Obviously, any bacterium, archaea, fungus, parasite, or virus that is introduced into a sample analyzed after the death of an individual, and any fragment of such a microorganism, nucleic acid, or other biomolecule, can be regarded as a contaminant. Moreover, contamination of an ancient specimen by ancient surrounding flora is a common taphonomical process. For example, it has been shown that the postmortem translocation of gut flora to other tissues is involved in evolution and taphonomical tissue calcification and preservation. This natural flora contamination of the environment in which an individual is placed is difficult to date. It can be prevented when the study sample was embedded within a hermetic receptacle. This was the case during our paleobacteriology and paleovirology investigations of a coprolite buried in the Middle Ages in a closed barrel that was used at that time as a latrine (3, 4).

ANTHROPOLOGICAL CONTAMINATION

Contamination of an old specimen during its discovery and storage can and must be prevented. Basically, this involves anthropological contamination; microorganisms can be

transmitted from operators to a specimen from the hands, from the scalp, and from the air. This contamination can be prevented by wearing gloves and gowning the entire body, especially the scalp, as is standard practice for preventing contamination by human DNA from operators during criminology operations. Next, specimens must be stored within suitable sterile containers, and hot, humid storage places that may precipitate aDNA decay must be avoided. Specimens must then be manipulated in microbiological terms— i.e., under a laminar flow hood while gloves and clean laboratory clothing are worn.

PREVENTION OF CONTAMINATION IN THE MICROBIOLOGY LABORATORY

Ancient specimens can be contaminated during handling in the microbiology laboratory by flora from operators and by environmental flora, especially particular microorganisms studied in the same laboratory and nucleotide sequences previously amplified in the same laboratory (Fig. 1). A first line of prevention is compliance with the principle of forward progress— manipulating specimens from the least potentially contaminated room to a potentially heavily contaminated one; the different stages of sample handling take place in areas separate from one another, progressing from the sparsely contaminated steps to the steps with high density. In such a laboratory, it is appropriate to organize a separate circuit for handling ancient (and modern) specimens that contain a low density of microorganisms and microbial aDNA and a separate circuit for handling bacterial, archaeal, viral, and parasitic strains, which obviously contain high quantities of microbial aDNA. These principles can be maximized by manipulating samples in a clean laboratory that is reserved for this purpose and limits, through specific protocols, the risk for contamination, in particular by modern DNA. Similarly, it is possible to practice the various handling stages in different laboratories, as we previously

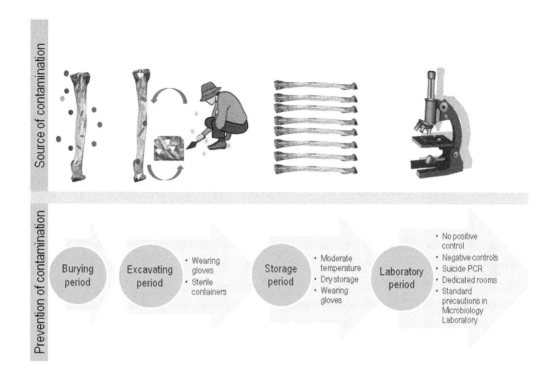

FIGURE 1 **Sources and prevention of contamination of ancient specimens by microbes and microbial products in the common example of buried specimens. Red bars indicate a pathogen present in the specimen at the time of death; blue and green circles indicate environmental microbes naturally contaminating the specimen during taphonomic processes; blue stars and yellow triangles indicate microbial contamination from discoverers of the specimen and fomites.**

described in ancient plague diagnosis (5). In the specific case of aDNA amplification by PCR, we have thus proposed an original protocol, "suicide PCR," in which a new pair of PCR primers is used for any amplification in the presence of a negative control and in the absence of any positive control; a new region of the genome is targeted for each new run of amplification with the idea of avoiding any possible contamination by the currently targeted region (5, 6). This approach is enabled by the current availability of complete genome sequencing for any pathogen of interest, which allows the easy design of a new, specific genomic region to be amplified for every new run of PCR. This approach has been successful and is currently used for the modern diagnosis of infections (7). Also, the complete withdrawal of any nested and seminested PCR protocol

and, conversely, the use of closed, real-time PCR protocols have sharply contributed to decreasing the risk for contamination during PCR analysis of aDNA. Indeed, nested PCR carried a very high risk for contamination by amplified DNA, when tubes from the first round of PCR had to be opened before the second (nested) round of PCR. We recommend that nested PCR no longer be used at all for the diagnosis of ancient (and modern) infections, and the authenticity of previous results based on nested PCR should be questioned, except those based on suicide PCR.

DETECTION OF CONTAMINATION

The basic element for detecting contamination from an old sample is the introduction of negative controls at each stage of manipulation of the old sample, during its direct or indirect observation, through the observation of biomolecules. The observation of differences between an old sample and the negative control is currently the basis for the interpretation of data. The choice of negative controls is a key point of paleomicrobiology analyses; the general principle is that the negative controls should be as close as possible to the old sample. There is no standard regarding the number of negative controls, but a proportion of one to five or one to ten specimens is reasonable, being in fact inversely proportional to the rarity of ancient specimens. As for biomolecule detection, negative controls must be introduced and manipulated strictly at the same time, in parallel with the ancient specimens, from the very beginning of manipulation of the ancient specimens, including, for example, not only PCR controls but also DNA extraction controls. This requirement is obviously modulated by the scarcity of the latter, especially in observations using destructive methods for the detection of biomolecules. However, additional negative controls may be introduced during handling to control key points of manipulation. This was the case in the manipulation using PCR, when the first negative controls had to be introduced from the step of extracting the nucleic acids, whereas the second negative controls were introduced during the preparation step of the PCR mixes. One particular negative control for buried specimens consists of sediments surrounding the buried specimen (8).

A second element that is rarely used in paleomicrobiology is the introduction of internal positive controls in the old specimen, to detect and interpret experimental negative results. In PCR-based analyses, positive controls are typically irrelevant synthetic oligonucleotide sequences and phage sequences. They are introduced just before nucleic acid extraction to test the efficiency of the extraction of nucleic acids, and just after extraction and just before PCR amplification to test the inhibition of PCR (9).

A third contamination-detecting element for analysis consists of biomolecules detected in the ancient sample. Regarding fragment detection of the 16S rDNA bacterial gene from ancient samples, re-sequence analysis showed that the low degree of sequence divergence of "old" with contemporary sequences was not compatible with dating and indicated that therefore the "old" sequences were actually the result of contamination of the old sample by contemporary sequences. This type of analysis is of course possible for biomolecules that we think have an evolutionary scale and for compatible time scales, which is very rarely the case.

A number of chemical and physical changes in aDNA molecules have been described for authenticating the "old" character of the sequences obtained. The most regularly observed changes concern the chemical deamination of the purine base adenosine, more pronounced at the end of molecules; with greater antiquity of a sample, a higher ratio of deamination is expected (10). Likewise, fragmentation of DNA molecules into fragments smaller than 200 bp has been linked to depurination (10). Moreover, various theories predicted the fragmentation of ancient DNA into fragments of less than 100 bp (11–13). It was concluded from theoretical postulates that the detection of larger fragments of aDNA would not be possible and would necessarily result from contamination. It must be emphasized, however, that most of these observations have been made with human aDNA and mammal aDNA in general and may not be directly extrapolated to any microbial aDNA (14). Interestingly, a recent study of extinct bone aDNA showed that mitochondrial aDNA (here assimilated to microbial aDNA) degraded twice as slowly as nuclear aDNA, with a great disparity in aDNA fragment length among the various specimens (15). In particular, aDNA repair has been observed in ancient microbes (16). We previously showed in an experimental model that modern mycobacterial DNA was more resistant to fragmentation than was modern mammal DNA (17). Further review of the published size of bacterial aDNA fragments indicates a wide range of lengths, with small fragments present along with larger ones in the same set of specimens (Fig. 2). It must be noted that the metagenomics of aDNA incorporates an experimental step of DNA shearing that renders the actual sizing of aDNA impossible.

DATA INTERPRETATION

In paleomicrobiology, the demonstrative weight of negative results is lower than that of positive results. For example, the failure to detect specific sequences of *Yersinia pestis* in human samples collected in a London plague cemetery has been interpreted as evidence supporting the hypothesis that this historic epidemic, and contemporary continental epidemics, were not plague epidemics (18, 19), and that the positive results obtained previously by our team in other mass graves were the result of contamination (5, 20). A few years after this negative work, some of the authors published the complete sequence of the genome of *Y. pestis*, reconstituted from the same mass grave in London that had been

identified as negative a few years earlier (21, 22). The same authors also questioned the detection of *Plasmodium* in the mummy Tutankhamun (23). As stated above, the introduction of internal controls is possible in order to mitigate the risk for over-interpreting negative results that can be linked, for example, by inhibition of PCR reactions by poorly characterized constituents contained in the ancient samples. The second element of interpretation is, of course, the absence of detection of the microorganism or biomolecule in the negative controls.

A third element in interpretation is the quantification of morphological and molecular changes attributable to taphonomical alterations compared with what is observed in contemporary samples of the same nature. As with microscopic observations of parasites in old samples, the expectation is to observe changes of structure or size in the ancient sample in comparison with modern structures. Regarding the analysis of microbial aDNA, a recent discovery of Bronze Age *Y. pestis* genomes incorporated the comparison between changes observed in microbial nucleotide sequences and the sequences of

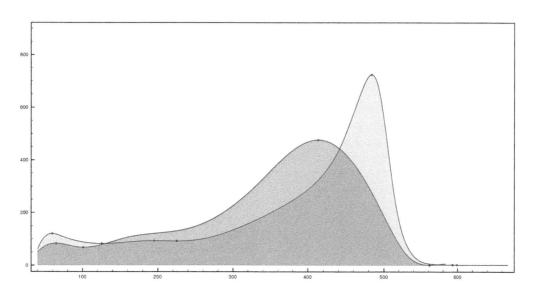

FIGURE 2 **Distribution of read length after throughput pyrosequencing of DNA extracted from a 14th century coprolite (blue curve) and from 1720 plague dental pulp specimens (red curve).**

the host to confirm the contemporary nature of the two sets of sequences, and thus the authenticity of the microbial sequences. This is based on the hypothesis that *Y. pestis* aDNA and the human host had been exposed to similar environmental conditions during the same period.

A fourth element is the observation in the same sample of molecules of a different nature yielding concordant identifications (24). Concerning nucleic acids, it is possible to detect nonspecificity by simple staining, such as with acridine orange. Similarly, we have shown the presence of DNA in a coprolite in which PCR-based analyses specified the sequences (3). Also, detection by the fluorescence *in situ* hybridization (FISH) method of DNA sequences detected by another method based on PCR or metagenomics authenticates molecular data. We have used this strategy for the detection of the methanogenic archaeal organism *Methanobrevibacter massiliense* in old samples of dental plaque (25). This approach has also been developed for the diagnosis of old cases of tuberculosis (25) and leprosy (26), in which molecular diagnosis by sequencing aDNA was supported by the analysis of mycolic acids, including in old cattle samples (27). Finally, the observation of more than two identifying sequences by techniques based on PCR or more independent sequences by identifying metagenomics is a key criterion for the authentication of molecular data (28). With plague, for example, it is possible to obtain cross-detection of the pathogen by detecting both specific aDNA sequences and a specific F1 antigen in the same ancient specimen, or in the same set of specimens from the same burial site (29–31).

Although the fact that only some samples in a batch are positive has sometimes been used as an argument to authenticate results (32), this argument should in fact not be maintained, the contaminations being only exceptionally systematic and more often few and seemingly random (Table 1).

PREVENTION OF CONTAMINATION

It is obviously not possible to prevent the contamination of natural old samples by changing the flora in contact with them during the burial period. However, natural contamination may be less important for samples preserved in relatively tight containers at the time of disposal. For example, we were able to analyze a medieval coprolite found in a barrel that served as a latrine and to reasonably interpret the viruses and bacteria found in this sample as authentic (3, 4). At the time of discovery of the old material, it is desirable to manipulate it by introducing physical barriers between the old specimen and the operators. We recommend wearing nonsterile gloves at a minimum, but clothing barriers can be completed by wearing a mask, a cap, or a complete gown.

These precautions limit the risk for DNA contamination by operators in the context of aDNA analysis of an individual, and for microbial contamination, including hand-borne contamination from the operators themselves and their environment. An ancient specimen should be placed in a suitable sterile, DNA-free receptacle. Storage must be carried out in conditions of temperature and humidity mimicking the natural conditions in which the old specimen was discovered.

Several recommendations have gradually been developed for handling old samples in the laboratory (33). A first key point is not to

TABLE 1 A list of criteria for the authentication of microbes in ancient specimens appears below

- Laboratory expertise in microbial diagnosis
- Adherence to standards for the prevention of contamination in the laboratory
- Absence of nested PCR
- Absence of positive control over the entire process of data production
- Negativity of the negative controls at every step of data production
- Suicide PCR
- Two independent aDNA sequences
- One aDNA sequence plus one non-DNA biomolecule

introduce any positive control–i.e., a sample of the same nature as the one sought in the old sample. It is always possible to validate a PCR-based protocol in a laboratory other than the one in which ancient specimens have to be manipulated. In the field of paleopathology— while detecting human biomolecules, for example—manipulation in a clean room by a fully gowned staff so as to establish barriers between operators and previous samples is the rule (34). In the field of paleomicrobiology, the use of clean rooms for handling old samples is not routine.

CONCLUSION

The description of microbial flora (bacteria, viruses, fungi, and parasites) in ancient specimens and the identification of ancient pathogens, collectively described by the term *paleomicrobiology*, is the same as microbiological identification in modern samples. As in all diagnostic areas, the quality of the results essentially depends on the experience of those conducting experiments, as demonstrated by the number and variety of tests they practice routinely. From this point of view, the foreseeable evolution in paleomicrobiology is to integrate more and more laboratories into diagnostic laboratories with practice and expertise in infectious disease or microbiota data exploration. The development of new diagnostic techniques such as paleoproteomics is a research area, as well as the interpretation of data derived from routine paleomicrobiology, from the perspective of acquiring better knowledge and understanding of the epidemiology of past and modern infectious diseases, as we have gradually done in the model of the plague.

CITATION

Drancourt M. 2016. Paleomicrobiology data: authentification and interpretation. Microbiol Spectrum 4(3):PoH-0017-2015.

REFERENCES

1. **Drancourt M, Raoult D.** 2005. Palaeomicrobiology: current issues and perspectives. *Nat Rev Microbiol* **3**:23–35.

2. **Huynh HT, Verneau J, Levasseur A, Drancourt M, Aboudharam G.** 2015. Bacteria and archaea paleomicrobiology of the dental calculus: a review. *Mol Oral Microbiol.* doi:10.1111/omi.12118.

3. **Appelt S, Francello L, Le Bailly M, Raoult D, Drancourt M, Desnues C.** 2014. Viruses in a 14th-century coprolite. *Appl Environ Microbiol* **80**:2648–2655.

4. **Appelt S, Armougom F, Le Bailly M, Robert C, Drancourt M.** 2014. Polyphasic analysis of a middle ages coprolite microbiota, Belgium. *PLoS One* **9**:e88376. doi:10.1371/journal.pone.0088376.

5. **Raoult D, Aboudharam G, Crubézy E, Larrouy G, Ludes B, Drancourt M.** 2000. Molecular identification by "suicide PCR" of *Yersinia pestis* as the agent of medieval black death. *Proc Natl Acad Sci U S A* **97**:12800–12803.

6. **Wiechmann I, Grupe G.** 2005. Detection of *Yersinia pestis* DNA in two early medieval skeletal finds from Aschheim (Upper Bavaria, 6th century A.D.). *Am J Phys Anthropol* **126**:48–55.

7. **Fournier PE, Raoult D.** 2004. Suicide PCR on skin biopsy specimens for diagnosis of rickettsioses. *J Clin Microbiol* **42**:3428–3434.

8. **Thèves C, Senescau A, Vanin S, Keyser C, Ricaut FX, Alekseev AN, Dabernat H, Ludes B, Fabre R, Crubézy E.** 2011. Molecular identification of bacteria by total sequence screening: determining the cause of death in ancient human subjects. *PLoS One* **6**:e21733. doi:10.1371/journal.pone.0021733.

9. **Ninove L, Nougairede A, Gazin C, Thirion L, Delogu I, Zandotti C, Charrel RN, De Lamballerie X.** 2011. RNA and DNA bacteriophages as molecular diagnosis controls in clinical virology: a comprehensive study of more than 45,000 routine PCR tests. *PLoS One* **6**:e16142. doi:10.1371/journal.pone.0016142.

10. **Dabney J, Meyer M, Pääbo S.** 2013. Ancient DNA damage. *Cold Spring Harb Perspect Biol* **5** pii: a012567.

11. **Lindahl T.** 1993. Recovery of antediluvian DNA. *Nature* **365**:700.

12. **Poinar HN, Höss M, Bada JL, Pääbo S.** 1996. Amino acid racemization and the preservation of ancient DNA. *Science* **272**:864–866.

13. **Smith CI, Chamberlain AT, Riley MS, Cooper A, Stringer CB, Collins MJ.** 2001. Neanderthal DNA. Not just old but old and cold? *Nature* **410**:771–772.

14. **Gilbert MTP, Willerslev E, Hansen EA, Barnes I, Rudbeck L, Lynnerup N, Cooper A.** 2003. Distribution patterns of postmortem

damage in human mitochondrial DNA. *Am J Hum Genet* **72**:32–47.

15. **Allentoft ME, Collins M, Harker D, Haile J, Oskam CL, Hale ML, Campos PF, Samaniego JA, Gilbert MT, Willerslev E, Zhang G, Scofield RP, Holdaway RN, Bunce M.** 2012. The half-life of DNA in bone: measuring decay kinetics in 158 dated fossils. *Proc Biol Sci* **279:** 4724–4733.

16. **Johnson SS, Hebsgaard MB, Christensen TR, Mastepanov M, Nielsen R, Munch K, Brand T, Gilbert MT, Zuber MT, Bunce M, Rønn R, Gilichinsky D, Froese D, Willerslev E.** 2007. Ancient bacteria show evidence of DNA repair. *Proc Natl Acad Sci U S A* **104**:14401–14405.

17. **Nguyen-Hieu T, Aboudharam G, Drancourt M.** 2012. Heat degradation of eukaryotic and bacterial DNA: an experimental model for paleomicrobiology. *BMC Res Notes* **5**:528.

18. **Prentice MB, Gilbert T, Cooper A.** 2004. Was the Black Death caused by *Yersinia pestis*? *Lancet Infect Dis* **4**:72.

19. **Gilbert MT, Cuccui J, White W, Lynnerup N, Titball RW, Cooper A, Prentice MB.** 2004. Absence of *Yersinia pestis* specific DNA in human teeth from five European excavations of putative plague victims. *Microbiology* **150**:341–354.

20. **Drancourt M, Aboudharam G, Signoli M, Dutour O, Raoult D.** 1998. Detection of 400-year-old *Yersinia pestis* DNA in human dental pulp: an approach to the diagnosis of ancient septicemia. *Proc Natl Acad Sci U S A* **95**:12637–12640.

21. **Bos KI, Schuenemann VJ, Golding GB, Burbano HA, Waglechner N, Coombes BK, McPhee JB, DeWitte SN, Meyer M, Schmedes S, Wood J, Earn DJ, Herring DA, Bauer P, Poinar HN, Krause J.** 2011. A draft genome of *Yersinia pestis* from victims of the Black Death. *Nature* **478**:506–510.

22. **Schuenemann VJ, Bos K, DeWitte S, Schmedes S, Jamieson J, Mittnik A, Forrest S, Coombes BK, Wood JW, Earn DJ, White W, Krause J, Poinar HN.** 2011. Targeted enrichment of ancient pathogens yielding the pPCP1 plasmid of *Yersinia pestis* from victims of the Black Death. *Proc Natl Acad Sci U S A* **10**:E746–E752. doi:10.1073/pnas.1105107108.

23. **Hawass Z, Gad YZ, Ismail S, Khairat R, Fathalla D, Hasan N, Ahmed A, Elleithy H, Ball M, Gaballah F, Wasef S, Fateen M, Amer H, Gostner P, Selim A, Zink A, Pusch CM.** 2010. Ancestry and pathology in King Tutankhamun's family. *JAMA* **303**:638–647.

24. **Tran TN, Aboudharam G, Raoult D, Drancourt M.** 2011. Beyond ancient microbial DNA: nonnucleotidic biomolecules for paleomicrobiology. *Biotechniques* **50**:370–380.

25. **Lee OY, Wu HH, Besra GS, Rothschild BM, Spigelman M, Hershkovitz I, Bar-Gal GK, Donoghue HD, Minnikin DE.** 2015. Lipid biomarkers provide evolutionary signposts for the oldest known cases of tuberculosis. *Tuberculosis (Edinb)* **95**(Suppl 1):S127–S132.

26. **Minnikin DE, Lee OY, Wu HH, Besra GS, Bhatt A, Nataraj V, Rothschild BM, Spigelman M, Donoghue HD.** 2015. Ancient mycobacterial lipids: key reference biomarkers in charting the evolution of tuberculosis. *Tuberculosis (Edinb)* **95** (Suppl 1):S133–S139.

27. **Lee OY, Wu HH, Donoghue HD, Spigelman M, Greenblatt CL, Bull ID, Rothschild BM, Martin LD, Minnikin DE, Besra GS.** 2012. *Mycobacterium tuberculosis* complex lipid virulence factors preserved in the 17,000-year-old skeleton of an extinct bison, *Bison antiquus*. *PLoS One* **7**:e41923. doi:10.1371/journal.pone. 0041923.

28. **Witas HW, Donoghue HD, Kubiak D, Lewandowska M, Głądykowska-Rzeczycka JJ.** 2015. Molecular studies on ancient *M. tuberculosis* and *M. leprae*: methods of pathogen and host DNA analysis. *Eur J Clin Microbiol Infect Dis* **34**:1733–1749.

29. **Bianucci R, Rahalison L, Massa ER, Peluso A, Ferroglio E, Signoli M.** 2008. Technical note: a rapid diagnostic test detects plague in ancient human remains: an example of the interaction between archeological and biological approaches (southeastern France, 16th-18th centuries). *Am J Phys Anthropol* **136**:361–367.

30. **Malou N, Tran TN, Nappez C, Signoli M, Le Forestier C, Castex D, Drancourt M, Raoult D.** 2012. Immuno-PCR—a new tool for paleomicrobiology: the plague paradigm. *PLoS One* **7**:e31744. doi:10.1371/journal.pone.0031744.

31. **Huynh HT, Gotthard G, Terras J, Aboudharam G, Drancourt M, Chabrière E.** 2015. Surface plasmon resonance imaging of pathogens: the *Yersinia pestis* paradigm. *BMC Res Notes* **8**:259.

32. **Rasmussen S, Allentoft ME, Nielsen K, Orlando L, Sikora M, Sjögren KG, Pedersen AG, Schubert M, Van Dam A, Kapel CM, Nielsen HB, Brunak S, Avetisyan P, Epimakhov A, Khalyapin MV, Gnuni A, Kriiska A, Lasak I, Metspalu M, Moiseyev V, Gromov A, Pokutta D, Saag L, Varul L, Yepiskoposyan L, Sicheritz-Pontén T, Foley RA, Lahr MM, Nielsen R, Kristiansen K, Willerslev E.** 2015. Early divergent strains of *Yersinia pestis* in Eurasia 5,000 years ago. *Cell* **163**:571–582.

33. **Willerslev E, Cooper A.** 2005. Ancient DNA. *Proc Biol Sci* **272**:3–16.

34. **Audic S, Béraud-Colomb E.** 1997. Ancient DNA is thirteen years old. *Nat Biotechnol* **15**:855–858.

Human Coprolites as a Source for Paleomicrobiology

7

SANDRA APPELT,[1] MICHEL DRANCOURT,[1] and MATTHIEU LE BAILLY[2]

Paleomicrobiology—the search for ancient microbes—is based on the analysis of human bones, teeth, and mummified soft tissues (1–3). In addition, ancient fecal remains preserved by mineralization or desiccation in the form of organic or permineralized coprolites, as well as intestinal contents and latrines, have yielded data on the environmental and gut microbiota of humans and animals that lived centuries to millennia ago (4–10). Also, coprolites appear to be a valuable source to study the diet of past populations (11). Combined, microbial and diet data support hypotheses regarding the role of environmental and cultural factors in the evolution of the human gut microbiota.

The first-ever described coprolite was discovered in the abdominal cavity of an ichthyosaur that was 230 million years old (4). The oldest studied human coprolite was dated to approximately 12,450 before the present (BP) (12). Since 1992, human coprolites have been reported from at least 87 archeological sites (Fig. 1). These coprolites were found in rock shelters, archeological layers, latrines, and pits, and inside mummified bodies (13–16). Human coprolites are usually fragmented, preserved as amorphous masses of different sizes and textures (16).

Consequently, they are often collected during the initial screening processes used to separate dirt from artifacts (16). Here, we review the techniques

[1]Aix Marseille Université, URMITE, UM63, CNRS 7278, IRD 198, Inserm 1095, Marseille, France; [2]Franche-Comté University, CNRS UMR 6249 Chrono-Environment, Besançon, France.
Paleomicrobiology of Humans
Edited by Michel Drancourt and Didier Raoult
© 2016 American Society for Microbiology, Washington, DC
doi:10.1128/microbiolspec.PoH-0002-2014

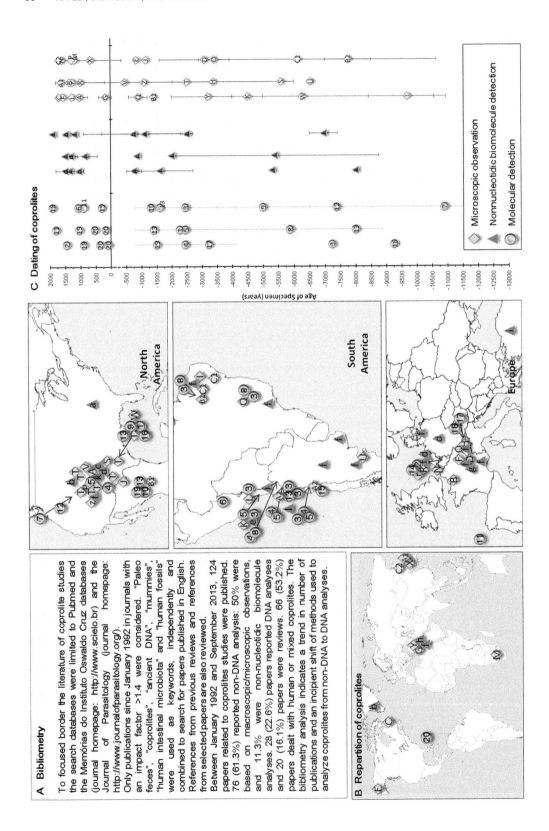

A Bibliometry

To focused border the literature of coprolite studies the search databases were limited to Pubmed and the Memórias do Instituto Oswaldo Cruz databases (journal homepage: http://www.scielo.br) and the Journal of Parasitology (journal homepage: http://www.journalofparasitology.org/).

Only publications since January 1992 in journals with an impact factor >1.4 were considered. "Paleo feces", "coprolites", "ancient DNA", "mummies", "human intestinal microbiota" and "human fossils" were used as keywords, independently and combined to search for papers published in English. References from previous reviews and references from selected papers are also reviewed.

Between January 1992 and September 2013, 124 papers related to coprolites studies were published. 76 (61.3%) reported non-DNA analysis: 50% were based on macroscopic/microscopic observations, and 11.3% were non-nucleotidic biomolecule analyses. 28 (22.6%) papers reported DNA analyses and 20 (16.1%) papers were reviews. 66 (53.2%) papers dealt with human or mixed coprolites. The bibliometry analysis indicates a trend in number of publications and an incipient shift of methods used to analyze coprolites from non-DNA to DNA analyses.

B Repartition of coprolites

C Dating of coprolites

Age of Specimen (years)

Microscopic observation
Nonnucleotidic biomolecule detection
Molecular detection

used to handle coprolites and outline knowledge about ancient gut microbiota as well as systemic infections that have been obtained by the analysis of human coprolites.

LOOKING FOR MICROORGANISMS IN COPROLITES: METHODOLOGY

During the excavation, the use of gloves is recommended to minimize contamination (6), especially when ancient DNA analyses are expected. Further, taphonomic observations of the archeological site can be of major interest, providing information about the removal by the action of water of microscopic remains and DNA during the fossilization process. For instance, the accumulation of coprolites or struvite (a soft mineral that precipitates in alkaline urine and forms stones) is a taphonomic process that occurs if fecal deposits did not have permanent protection from environmental conditions throughout the centuries (6, 12, 17–21). After excavation, the specimen can be stored in sterile forensic bags to avoid contamination, fungal growth, and further degradation. Changes in environmental conditions (light, temperature, oxygen concentration) from the place of specimen discovery can induce microbial growth or accelerate ancient DNA (aDNA) degradation (6, 22). The coprolite should be photographed, measured, and weighted before it is archived in the laboratory. Morphometric data on form, size, color, texture, and inclusions can indicate what animal produced the coprolite (23). Nevertheless, with coprolites, host identification is more often based on the archeological context, host-specific food residues, host-specific parasites, and bacteria

found inside the coprolite and on the analysis of host aDNA (12, 24–27).

Before coprolites are analyzed, it is important to brush or remove the external layer aseptically. This is to prevent subaerial exposure of the coprolite surface to contaminants. Also, the transport and storage of many coprolites in a single bag can result in cross-contamination between coprolites (8, 10, 28). A rehydration phase is needed and can be extended up to 10 days (29–31). Rehydration is done in aqueous 0.5% trisodium phosphate solution, sometimes completed with 5% glycerinated water or several drops of 10% formalin to avoid fungal or bacterial growth (14, 31–33). When the rehydration is performed in water only, sodium hydroxide, or ethylenediamine-tetraacetic acid (EDTA), biogenic components can be destroyed and the coprolite can disintegrate (14). Accordingly, macromolecules including DNA, proteins, and lipids can be stained with acridine orange, Fast Green FCF, and Nile red, respectively, before analysis (10).

Microscopic Analysis

Microscopic observations can reveal helminths eggs and protozoa cysts (14, 34). These microbial residues can remain uncollapsed for at least 11,000 years inside human coprolites (Fig. 1). Autofluorescence was described for the intestinal parasite *Cyclospora cayetanensis* that might have been maintained for at least 3,000 years (35). The identification of intestinal parasites and protozoa is based on morphometric characterization of their eggs and cysts with light microscopy. For instance, length and width differentiate the human-infecting *Trichuris*

FIGURE 1 **(A) Bibliometry. (B) Repartition and dating of human and mixed coprolites investigated during the last 21 years. The human coprolites reviewed herein were found in approximately 87 different archeological sites: 37 (42.5%) and 24 (27.6%) were located in South America and North America respectively, 18 (20.7%) were found in Europe, and 3 to 5 (3.4% to 5.7%) were found in Africa and Asia. Macroscopic / microscopic analyses are labeled with yellow squares, non-nucleotidic biomolecular analyses are with blue triangles, and molecularly analyzed coprolites with green circles. The corresponding references are listed in Table 2. (C) Dating of coprolites.**

trichiura eggs and the pig-infecting *Trichuris suis* eggs (26, 36, 37). Furthermore, the eggs of *Metagonimus* spp. are differentiated from those of *Clonorchisis* spp. only by the scanning electron microscopic observation of differences on their surface structures (38). The observation and identification of protozoa cysts are rare because of reduced resistance to natural decay and the rehydration protocol (39, 40).

Non-nucleotidic Biomolecule Detection

Cytochemical staining with Fast Green detected proteins in human coprolites dated to between 100 BP and 600 AD (10). It was further possible to detect intestinal protozoa antigens by using either commercially available immunofluorescence assay (IFA) or enzyme-linked immunosorbent assay (ELISA) (32, 34, 35, 39–42). A monoclonal antibody IFA test detected *Cryptosporidium* spp. and *Giardia* spp. in coprolites dated to between 3,000 BP and 1,000 AD (34, 35, 40). The ELISA detected an *Entamoeba histolytica*–specific adhesin as well as a *Giardia intestinalis*–specific cyst wall protein 1 (CWP1) and a glycoprotein (GSA65) in samples dated from 5,300 BP to 1,900 AD (32, 39, 40, 42).

Molecular Microbiology

Like all ancient materials, coprolites needed to be manipulated according to standard protocols for aDNA. This implies never using positive controls but instead incorporating several negative controls. Positive results require confirmation by additional tests: detection of another, unrelated aDNA sequence; detection of a specific protein; or the independent replication of results in an appropriated second laboratory (43–47).

aDNA Extraction

Recently, acridine orange staining detected aDNA in human coprolites dated to between 100 BP and 600 AD, thus indicating that molecular biological techniques can be applied to the specimens tested (10). Nevertheless, poorly characterized inhibitors may hamper the PCR-based detection of aDNA. Previous investigations showed that coprolites contain lignin and Maillard reaction products (cross-links between reducing sugars and amino groups), all inhibitors of PCR (33). Those inhibitors have to be removed before molecular tests are applied. N-Phenacyl thiazolium bromide in sodium phosphate buffer solution was described to be effective against Maillard reaction products (33, 48). Incorporation of internal controls such as artificial nontargeted DNA templates allows monitoring for PCR inhibitors (49–51). DNA extractions were performed either on eggs of intestinal parasites and bones pulled out from coprolites or on the coprolites directly (24, 33, 52). Until now, six different protocols have been reported for the successful extraction of aDNA from coprolites. An initial protocol described in 1998 (33) was then slightly modified depending on the material (25, 28, 53–55). The MoBio Power Soil and UltraClean Fecal DNA Kit and a salting-out extraction protocol were also used for coprolites dated from 7,315 BP to 1,450 AD. The latter is especially suitable for coprolites containing bone fragments (8, 22, 10, 56). In 2012, another protocol, normally used for ancient sediments, was used for human coprolites dated to 12,265 BP (12, 57). Finally, an extraction protocol specifically for coprolites was described in 2012 (58).

PCR Amplification and Sanger Sequencing

Intestinal parasites, fungi, bacteria, and archaea have been identified by sequencing PCR-amplified cytochrome b gene, 18S rDNA, 16S rDNA, and 5S rDNA, or nuclear ribosomal internal transcribed spacer (ITS) regions (Table 1). Following amplification, species are identified by blast annotations after sequencing or by terminal restriction fragment length polymorphisms (10, 15, 24, 28, 53–55, 61–64). Because nested PCR is affected by amplicon carryover during the second round of amplification and yields false-positive results (65–68), it should no longer be used in coprolite investigations.

Next-Generation Sequencing

Large-scale sequencing yields a wide range of DNA sequences simultaneously, either dependent on or independent of targeted gene amplification (69–71). In a first strategy, PCR amplification of variable regions (V3, V6) of bacterial 16S rDNA with universal primers is followed by massive sequencing (72). This approach yields data on the bacterial communities in coprolites (8). Moreover, additional source tracking performed on 16S rDNA data sets makes it possible to compare the mixture of bacterial taxa associated with coprolites with different modern sources (8). The same strategy incorporating 18S rDNA primers has been used to look for fungal DNA (73).

The second strategy is the massive sequencing of the DNA without previous amplification (68, 69). Whole-genome shotgun (WGS) sequencing strategies can be used to gain access to the collection of genomes associated with specimens. These include sequences associated with the host, eukaryotes, bacteria, and archaea, as shown for two 1,300-year-old coprolites (22, 69). Moreover, WGS data sets can be used to learn more about the major metabolic pathways associated with the collection of genomes associated with coprolites (22, 69).

OUTPUTS

Resident Gut Microbiota

The analysis of microbial communities preserved in coprolites, with the use of massive parallel sequencing or PCR amplifications, has indicated that parts of the ancient gut microbiota are preserved (8, 15, 22, 55, 61, 62) (Table 2). The seven phyla mainly found in the modern human gut flora—i.e., *Firmicutes*, *Bacteroidetes*, *Actinobacteria*, *Fusobacteria*, *Proteobacteria*, *Verrucomicrobia*, and *Cyanobacteria*—have all been detected in coprolites (8, 15, 22, 55, 61, 62). Source tracking performed on several pre-Columbian coprolites found a mixture of bacterial taxa similar to those coming from the stool of human children, primate gut, or compost, and in some cases no similarities to known sources were found (8). Moreover, recent analysis has identified *Methanobrevibacter* spp., *Methanosphaera* spp., and *Sulfolobus* spp. in coprolites (10, 22). Indeed, *Methanobrevibacter smithii* has been shown to be constant inhabitant of the human gut (106).

The comparison of human coprolite samples from two different Indonesian cultures indicated that the human intestinal microbiome was affected by diet related to cultural traditions (10). Furthermore, comparison with modern stool samples showed that coprolites exhibited more similarities to one another and to stools from rural communities than to stools from cosmopolitan communities. These findings suggest that the modern lifestyle may results in changes in the composition of the human gut flora (8, 22).

Pathogens

Microbiological analysis of modern feces has made possible the diagnosis of both intestinal and systemic infections (107–110). By analogy, coprolites are a material of choice for the retrospective diagnosis of diseases (Fig. 2).

Intestinal tract pathogens—helminths and protozoa—have been diagnosed thanks to the microscopic observation of eggs and cysts and to the detection of specific antigens and DNA. *Ascaris* spp. and *Trichuris* spp., causing ascariasis and trichuriasis, respectively, have been detected in 80% of human coprolites in European archeological sites and in 100% of those from medieval times (13, 14). Because

TABLE 1 PCR systems used to amplify microbial ancient DNA out of coprolites[a]

Microorganisms	Target	Method	Microbial aDNA, bp	Primer	Sequence 5'→3'	Positive-test specimen age	References
Eukaryotes							
Ascaris spp.	18S rDNA[b]	PCR	123	Asc6 / Asc7	CGAACGGCTCATTACAACAG / TCTAATAGATGCGCTCGTC	1,392–1,800 AD	24, 59
		PCR	99	Asc8 / Asc9	ATACATGCACCAAAGCTCCG / GCTATAGTTATTCAGAGTCACC		
		PCR	147	Asc10 / Asc11	CCATGCATGTCTAAGTTCAA / CARAAAWTCGGAGCTTTGGT		
	Cytochrome[b]	PCR	98–142	Asc1 / Asc2	GTTAGGTTACCGTCTAGTAAGG / CACTCAAAAAGGCCAAAGCACC	8,860 BP–1,905 AD	24, 60
	Cytochrome[c] oxidase subunit 1[c]	PCR	248	COX1F / As-Co1R	GGATCTTGACTCTCGGGCTTA / ACATAATGAAAATGACTAACAAC	1,800 AD	59
		PCR	199	As-Co1F / COX1R	TTTTTGGTCATCCTGAGGTTTAT / GCCCGAGAGTCAAGATCCAT		
	NADH dehydrogenase[c]	PCR	152	NAD1F / NAD1R	CTCCTCGAATTCTTCGGAAA / CAGAAAACCCAATCAAACACA	1,800 AD	59
Enterobius spp.	5S rDNA[d]	Nested PCR	419	Entf / Entr	CACTTGCTATACCAACAACAC / GCGCTACTAAACCATAGAG	4,110 BP–900 AD	54, 28
			198	Eva / Evb	ACAACACTTGCACGTCTC / GAATTGCTCGTTTGC		
Diverse fungi	ITS region	PCR	645	ITS1F / ITS4B	CTTGGTCATTTAGAGGAAGTAA / TCCTCCGCTTATTGATATGC	100 BP–500 AD	10

Bacteria

Organism	Gene	Method	bp	Primer	Sequence	Age	Ref
Bacteroides spp.	16S rDNA	PCR	541	HF183F / Bac708R	ATCATGAGTTCACATGTCCG / CAATCGGGAGTTCTTCGTG	100 BP–500 AD	10
Bacteroides spp.	16S rDNA	PCR	150	BacCan / Bac708R	GGAGGCGCAGACGGGTTTT / CAATCGGGAGTTCTTCGTG	100 BP–500 AD	10
Diverse bacteria (16S)	16S rDNA	PCR	523	8dF / K2R	AGAGTTTGTTCMTGGCTCAG / GTATTACCGCGGCTGCTGG	100 BP–500 AD	10
Diverse bacteria (16S)	16S rDNA	PCR	98	29f / 98r	TGGCTCAGATTGAACGCTG / CCAGACATTACTCACCCGTCC	980–1,170 AD	61, 55
Diverse bacteria (16S)	16S rDNA	PCR	556 / 576 / 797	8F / 564R / 584R / 805R	AGCGTCAAACTTTTAAATTGAA / CCTGCGTGCGCTTTACGCCC / ACATCTGACTTAACAAACCG / TCGACATCGTTTACGGCGTG	3,350–3,100 BP	62
Diverse bacteria (16S)	16S rDNA V6	PCR	188	341F / 529R	CCTACGGGRSGCAGCAG / ACCGCGGCKGCTGGC	1,400–8,000 BP	8

Archaea

Organism	Gene	Method	bp	Primer	Sequence	Age	Ref
Diverse archaea (16S)	16S rDNA	PCR	915	Arch21F / Arch958R	TTCCGGTTGATCCYGCCGGA / YCCGGCGTTGAMTCCAATT	100 BP–500 AD	10

[a]Shown are the targeted microorganisms, genomic regions, length of amplification products (bp) as corresponding primer sets, and age of those coprolite specimens that yielded positive amplification results.

[b]Overlapping primer set amplifying totally 176 bp of the 18S rDNA gene.

[c]Primer set led to unspecific amplification.

[d]No 419-bp amplification product was obtained.

Abbreviations: aDNA, ancient DNA; bp, base pair; spp., species; NADH, reduced nicotinamide adenine dinucleotide; ITS, internal transcribed spacer.

TABLE 2 Scientific studies performed on human and potentially mixed animal and human coprolites[a]

Method of analysis	Sample source	Archeological site	Identifiers	References
Microscopic observations	Human	Canyon de Chelly, Arizona (USA)	A	74
Microscopic observations	Human	Chungcheongnam (Korea)	A1	75
Microscopic observations	Human	Huaricanga, Caballete (Peru)	A2	76
Microscopic observations	Human	Paris (France)	B	77
Microscopic observations	Human	Durango (Mexico)	B1	78
Microscopic observations	Human	Piauí (Brazil)	B2	79
Microscopic observations	Mixed	Patagonia (Chile)	B3	80
Microscopic observations	Mixed	Ferryland (Canada)	C	81
Microscopic observation	Human	Yongin (Korea)	C1	38
Microscopic observation	Human	São Raimundo Nonato, Piauí, and Pernambuco (Brazil)	C2	82
Microscopic observations	Human	Chalain Lake (France)	D	83
Microscopic observations	Human	Aleutian Islands (Alaska)	E	84
Microscopic observations	Human	Montbéliard (France)	F	17
Microscopic observations	Human	Dakhleh Oasis (Egypt)	G	85
Microscopic observations	Human	Arbon-Bleiche (Switzerland)	H	30
Microscopic observations	Mixed	Pernambuco (Brazil)	I	86
Macroscopic observations	Human	Chihuahuan Desert (Mexico)	J	87
Microscopic observations	Human	Chinchorro (Chile)	K	23
Microscopic observations	Human (suspected)	Lluta Valley (Chile)	L	88
Microscopic observations	Human	Greefswald (South Africa)	M	89
Microscopic observations	Human (suspected)	Raversijde (Belgium)	N	37
Microscopic observations	Human	Arbon Bleiche (Switzerland)	O	31
Microscopic observations	Human	Clen Canyon, Utah (USA)	P	90
Microscopic observations	Human	Minas Gerais (Brazil)	Q	91
Microscopic observations	Mixed	Namur (Belgium)	R	36
Microscopic observations	Mixed	Namur (Belgium)	r	26
Microscopic observations	Human	Hinds Cave, Texas (USA)	S	92
Microscopic observations	Human	Nevada, Texas, Utah (USA)	T	93
Microscopic observations	Mixed	Patagonia (Agentina)	U	94
Microscopic observations	Human	Antelope Cave, Arizona (USA)	V	95
Microscopic observations	Mixed	Hinds Cave, Texas (USA)	W	96
Microscopic observations	Human	Yongin (Korea)	X	38, 97
Microscopic observations	Human	El-Deir, Oasis of Kharga (Egypt)	Y	98
Microscopic observations	Mixed	Mojave County, Arizona (USA)	Z	99
Molecular analysis	Human	Hinds Cave, Texas (USA)	1	25
Molecular analysis	Mixed	Namur (Belgium)	2	24
Molecular analysis	Human	Azapa, Antofagasta, Caserones (Chile)	3	53
Molecular analysis	Human	Piauí, Unai (Brazil)	3	53
Molecular analysis	Human	Antelope House, Canyon de Chelly, Arizona (USA)	4	54
Molecular analysis	Human	Tulan, Caserones, Tiliviche (Chile)	4	54
Molecular analysis	Human	Antelope House, Canyon de Chelly, Arizona (USA)	5	28
Molecular analysis	Human	Tulan, Caserones, Tiliviche (Chile)	5	28
Molecular analysis	Human	Cuzco (Peru)	6	55
Molecular analysis	Human	Oregon (USA)	7	20
Molecular analysis	Human	Brazil, Peru, Chile	8	60
Molecular analysis	Human	Hinds Cave, Texas (USA)	9	16

(Continued on next page)

TABLE 2 Scientific studies performed on human and potentially mixed animal and human coprolites[a]
(Continued)

Method of analysis	Sample source	Archeological site	Identifiers	References
Molecular analysis	Human	Rio Zape (Mexico)	10	100
Molecular analysis	Human	Tenerife (Spain)	11	59
Molecular analysis	Mixed	Oregon (USA)	12	12
Molecular analysis	Human	Caserones (Chile)	13	8
Molecular analysis	Human	Hinds Cave, Texas (USA)	13	8
Molecular analysis	Human	Rio Zape (Mexico)	13	8
Molecular analysis	Human	Andean	15	61
Molecular analysis	Human	Hinds Cave, Texas (USA)	16	12
Molecular analysis	Human	Alps (Austria/Italy)	17	62
Molecular analysis	Human	Northern Italy	18	52
Molecular analysis	Human	Rio Zape (Mexico)	19	15
Molecular analysis	Human	Sorcé, Island of Vieques; Guayanilla (Puerto Rico)	20	10
Non-nucleotidic biomolecule analysis	Human	Kentucky (USA)	a	56
Non-nucleotidic biomolecule analysis	Human	West coast of Greenland, Nevada (USA)	b	101
Non-nucleotidic biomolecule analysis	Human	Cowboy Wash. Colorado (USA)	c	102
Non-nucleotidic biomolecule analysis	Mixed	USA, Germany, Belgium	d	103
Non-nucleotidic biomolecule analysis	Human	Los Gavilanes (Peru)	e	39
Non-nucleotidic biomolecule analysis	Human (suspected)	Argentina, Brazil, Chile, USA, France, Belgium, Switzerland, Germany, Sudan	f	34
Non-nucleotidic biomolecule analysis	Human	La Mothe (France)	g	32
Non-nucleotidic biomolecule analysis	Human (suspected)	City of Acre (Israel)	h	40
Non-nucleotidic biomolecule analysis	Mixed	Konja (Turkey)	i	42
Non-nucleotidic biomolecule analysis	Human	Atacama Desert (Chile)	j	104
Non-nucleotidic biomolecule analysis	Human	Utah (USA)	k	105

[a]The type of investigation, the archaeological site where the studied coprolites were found, and the nature of the host are reported, with the corresponding references and the identifiers in Fig. 1.

Ascaris spp. are also known to infect pigs, this observation call into question the initial source of transmission in connection with the origin of agriculture in Neolithic times (13, 14, 63).

Dicrocoelium dendriticum (responsible for bile duct infections), *Taenia* spp. (cysticercosis), and *Diphyllobothrium* spp. (diphyllobothriasis) have also been detected (Fig. 2). *Diphyllobothrium* is known to infect freshwater fish, and transmission to humans can occur through the consumption of raw or undercooked fish. *Diphyllobothrium* spp. eggs were observed in 75% of human coprolites from late Neolithic times in Europe, suggesting the importance of this parasitic infection at the time (31). Protozoa cysts belonging to

Entamoeba spp., *Giardia* spp., and *Cyclospora* spp. were also identified in coprolites from Europe and America (14, 34, 35). A *Schistosoma* spp. egg was recently observed in the contents of a 6,000-year-old pelvis in northern Syria (111).

Furthermore, unicellular eukaryotes and bacteria are associated with coprolites. The dilated, coprolite-rich colon of a Peruvian Inca mummy (1,400 AD) suggested that Chagas disease, caused by *Trypanosoma* spp., might have circulated during that time. This observation was further confirmed by PCR and by electron microscopic observations performed on soft-tissue samples from the same mummy (112). By considering amplification products of the 16S rDNA gene with 74% to 100%

Pathogens identified in coprolites

| Microorganisme | Methode of identification |

Digestive tract pathogens
Intestinal helminths
Ancylostomid spp.
Ascaris spp.*
Capillaria spp.
Dicrocoelium spp.
Dioctophymidae spp.
Diphyllobothrium spp.
Dipylidium caninum
Enterobius spp.
Fasciola spp.
Gymnophalloides seoi
Metagonimus yokogawai
Physaloptera spp.
Schistosoma spp.
Taenia spp.
Trichostrongylus spp.
Trichuris spp.*
Intestinal protozoa
Cryptosporidium spp.
Cyclospora spp.
Entamoeba spp.
Giardia spp.*
Isospora belli
Sarcocystis hominis
Unicellular eukaryotes and bacteria
Protozoa
Leishmania spp.
Plasmodium spp.
Systemic bacterial pathogens
Clostridium spp.
Haemophilus spp.
Mycobacterium spp.
Neisseria spp.
Shigella spp.
Yersinia spp.

* additionally detected in a potentially mixed animal and human coprolite

FIGURE 2 **Intestinal and systemic pathogens identified in human coprolites. The pathogens are grouped based on their taxonomic classification into intestinal helminths, intestinal protozoa, and unicellular eukaryotes and bacteria. The methods of identification that yielded positive results are marked in green, and negative tests are marked in gray. Microscopic observations were performed with light or electron microscopy. Non-nucleotidic biomolecular detection included testing with immunofluorescence assay (IFA) and enzyme-linked immunosorbent assay (ELISA), and molecular detection was performed with PCR amplification and next-generation sequencing.**

similarities, bacteria from the genera *Acinetobacter* spp., *Clostridium* spp., and *Haemophilus* spp. were identified in human coprolites (Fig. 2) (55, 61). Sequences related to *Shigella* spp., responsible for shigellosis, have also been

detected (5). Furthermore, metagenomic sequences of pathogenic microorganisms were found in WGS data sets from pre-Columbian coprolites, including *Neisseria* spp. (*Neisseria gonorrhoeae*), *Yersinia* spp. (*Yersinia enterocolitica*), *Mycobacterium* spp. (*Mycobacterium tuberculosis*), *Plasmodium* spp., and *Shigella* spp. (22).

CONCLUSIONS

Coprolite microbiology provides information regarding the human gut microbiota and diseases in past populations, after the strict enforcement of standard practices in paleomicrobiology. These paleomicrobiological standards, developed over the years, are essential precautions in order to avoid in-laboratory contamination and thus ensure the interpretability of data. Currently, there are no established standardized procedures widely performed systematically to analyze coprolites, but the techniques are the same as those used for routine diagnosis. Accordingly, investigations performed on coprolites have already helped to estimate ancient human gut microbiota composition and to expand knowledge about the intestinal parasites that circulated in ancient populations, as well as diet in past populations (113). When parallels are drawn with data obtained from modern stool samples, coprolites provide data about systemic infections. Yet, few data have been obtained about bacteria and archaea, and not at all about viruses, compared with those about intestinal parasites. These disparities remain open fields for further investigation that will benefit from technical improvements and certainly provide useful data.

CITATION

Appelt S, Drancourt M, Le Bailly M. 2016. Human coprolites as a source for paleomicrobiology. Microbiol Spectrum 4(3): PoH-0002-2014.

REFERENCES

1. **Taylor GM, Goyal M, Legge AJ, Shaw RJ, Young D.** 1999. Genotypic analysis of *Mycobacterium tuberculosis* from medieval human remains. *Microbiology* **145:**899–904. http://dx.doi.org/10.1099/13500872-145-4-899.

2. **Zink A, Haas CJ, Reischl U, Szeimies U, Nerlich AG.** 2001. Molecular analysis of skeletal tuberculosis in an ancient Egyptian population. *J Med Microbiol* **50:**355–366. http://dx.doi.org/10.1099/0022-1317-50-4-355.

3. **Tran TN, Aboudharam G, Raoult D, Drancourt M.** 2011. Beyond ancient microbial DNA: non-nucleotidic biomolecules for paleomicrobiology. *Biotechniques* **50:**370–380. http://dx.doi.org/10.2144/000113689.

4. **Reinhard KJ, Bryant VMJ.** 1992. Coprolite analysis: a biological perspective on archaeology, p 245–288. *In* Shiffer M (ed), *Advances in Archaeological Method and Theory.* University of Arizona Press, Tucson, AZ.

5. **Araújo A, Reinhard K, Bastos OM, Costa LC, Pirmez C, Iñiguez A, Vicente AC, Morel CM, Ferreira LF.** 1998. Paleoparasitology: perspectives with new techniques. *Rev Inst Med Trop Sao Paulo* **40:**371–376. http://dx.doi.org/10.1590/S0036-46651998000600006.

6. **Bouchet F, Guidon N, Dittmar K, Harter S, Ferreira LF, Chaves SM, Reinhard K, Araújo A.** 2003. Parasite remains in archaeological sites. *Mem Inst Oswaldo Cruz* **98**(Suppl 1):47–52. http://dx.doi.org/10.1590/S0074-02762003000900009.

7. **Dittmar K.** 2009. Old parasites for a new world: the future of paleoparasitological research. a review. *J Parasitol* **95:**365–371. http://dx.doi.org/10.1645/GE-1676.1.

8. **Tito RY, Knights D, Metcalf J, Obregon-Tito AJ, Cleeland L, Najar F, Roe B, Reinhard K, Sobolik K, Belknap S, Foster M, Spicer P, Knight R, Lewis CM Jr.** 2012. Insights from characterizing extinct human gut microbiomes. *PLoS One* **7:**e51146. http://dx.doi.org/10.1371/journal.pone.0051146.

9. **Dentzien-Dias PC, Poinar G Jr, de Figueiredo AE, Pacheco AC, Horn BL, Schultz CL.** 2013. Tapeworm eggs in a 270 million-year-old shark coprolite. *PLoS One* **8:**e55007. http://dx.doi.org/10.1371/journal.pone.0055007.

10. **Santiago-Rodriguez TM, Narganes-Storde YM, Chanlatte L, Crespo-Torres E, Toranzos GA, Jimenez-Flores R, Hamrick A, Cano RJ.** 2013. Microbial communities in pre-columbian coprolites. *PLoS One* **8:**e65191. http://dx.doi.org/10.1371/journal.pone.0065191.

11. **Sistiaga A, Mallol C, Galván B, Summons RE.** 2014. The Neanderthal meal: a new perspective using faecal biomarkers. *PLoS One* **9:**e101045. http://dx.doi.org/10.1371/journal.pone.0101045.

12. **Jenkins DL, Davis LG, Stafford TW Jr, Campos PF, Hockett B, Jones GT, Cummings LS, Yost C, Connolly TJ, Yohe RM II, Gibbons SC, Raghavan M, Rasmussen M, Paijmans JL, Hofreiter M, Kemp BM, Barta JL, Monroe C, Gilbert MT, Willerslev E.** 2012. Clovis age Western Stemmed projectile points and human coprolites at the Paisley Caves. *Science* **337:**223–228. http://dx.doi.org/10.1126/science.1218443.

13. **Bouchet F, Harter S, Le Bailly M.** 2003. The state of the art of paleoparasitological research in the Old World. *Mem Inst Oswaldo Cruz* **98**(Suppl 1):95–101. http://dx.doi.org/10.1590/S0074-02762003000900015.

14. **Gonçalves ML, Araújo A, Ferreira LF.** 2003. Human intestinal parasites in the past: new findings and a review. *Mem Inst Oswaldo Cruz* **98**(Suppl 1):103–118. http://dx.doi.org/10.1590/S0074-02762003000900016.

15. **Rollo F, Ermini L, Luciani S, Marota I, Olivieri C.** 2006. Studies on the preservation of the intestinal microbiota's DNA in human mummies from cold environments. *Med Secoli* **18:**725–740.

16. **Reinhard KJ, Bryant VM.** 2008. Pathoecology and the future of coprolite studies in bioarchaeology. *Natural Resources* **43:**205–224.

17. **Bouchet F, Harter S, Paicheler JC, Aráujo A, Ferreira LF.** 2002. First recovery of *Schistosoma mansoni* eggs from a latrine in Europe (15-16th centuries). *J Parasitol* **88:**404–405.

18. **Carrio J, Riquelme J, Navarro C, Munuera M.** 2001. Pollen in hyaena coprolites reflects late glacial landscape in southern Spain. *Palaeogeogr Palaeoclimatol Palaeoecol* **176:**193–205. http://dx.doi.org/10.1016/S0031-0182(01)00338-8.

19. **Chaves SDM, Reinhard K.** 2006. Critical analysis of coprolite evidence of medicinal plant use, Piauí, Brazil. *Palaeogeogr Palaeoclimatol Palaeoecol* **237:**110–118. http://dx.doi.org/10.1016/j.palaeo.2005.11.031.

20. **Gilbert MT, Jenkins DL, Götherstrom A, Naveran N, Sanchez JJ, Hofreiter M, Thomsen PF, Binladen J, Higham TF, Yohe RM II, Parr R, Cummings LS, Willerslev E.** 2008. DNA from pre-Clovis human coprolites in Oregon, North America. *Science* **320:**786–789. http://dx.doi.org/10.1126/science.1154116.

21. **Goldberg P, Berna F, Macphail RI.** 2009. Comment on "DNA from pre-Clovis human coprolites in Oregon, North America." *Science* **325:**148, author reply 148. http://dx.doi.org/10.1126/science.1167531.

22. **Tito RY, Macmil S, Wiley G, Najar F, Cleeland L, Qu C, Wang P, Romagne F, Leonard S, Ruiz**

AJ, Reinhard K, Roe BA, Lewis CM Jr. 2008. Phylotyping and functional analysis of two ancient human microbiomes. *PLoS One* **3**:e3703. http://dx.doi.org/10.1371/journal.pone.0003703.

23. Reinhard K, Fink TM, Skiles J. 2003. A case of megacolon in Rio Grande valley as a possible case of Chagas disease. *Mem Inst Oswaldo Cruz* **98**(Suppl 1):165–172. http://dx.doi.org/10.1590/S0074-02762003000900025.

24. Loreille O, Roumat E, Verneau O, Bouchet F, Hänni C. 2001. Ancient DNA from *Ascaris*: extraction amplification and sequences from eggs collected in coprolites. *Int J Parasitol* **31**:1101–1106. http://dx.doi.org/10.1016/S0020-7519(01)00214-4.

25. Poinar HN, Kuch M, Sobolik KD, Barnes I, Stankiewicz AB, Kuder T, Spaulding WG, Bryant VM, Cooper A, Pääbo S. 2001. A molecular analysis of dietary diversity for three archaic Native Americans. *Proc Natl Acad Sci U S A* **98**:4317–4322. http://dx.doi.org/10.1073/pnas.061014798.

26. da Rocha GC, Serra-Freire NM. 2009. Paleoparasitology at "Place d'Armes", Namur, Belgium: a biostatistics analysis of trichurid eggs between the Old and New World. *Rev Bras Parasitol Vet* **18**:70–74. http://dx.doi.org/10.4322/rbpv.01803013.

27. Fugassa MH, Gonzalez Olivera EA, Petrigh RS. 2013. First palaeoparasitological record of a dioctophymatid egg in an archaeological sample from Patagonia. *Acta Trop* **128**:175–177. http://dx.doi.org/10.1016/j.actatropica.2013.06.001.

28. Iñiguez AM, Reinhard K, Carvalho Gonçalves ML, Ferreira LF, Araújo A, Paulo Vicente AC. 2006. SL1 RNA gene recovery from *Enterobius vermicularis* ancient DNA in pre-Columbian human coprolites. *Int J Parasitol* **36**:1419–1425. http://dx.doi.org/10.1016/j.ijpara.2006.07.005.

29. Dufour B, Le Bailly M. Testing new parasite egg extraction methods in paleoparasitology and an attempt at quantification. *Int J Paleopathol*, in press. doi:10.1016/j.ijpp.2013.03.008.

30. Le Bailly M, Leuzinger U, Bouchet F. 2003. *Dioctophymidae* eggs in coprolites from Neolithic site of Arbon-Bleiche 3 (Switzerland). *J Parasitol* **89**:1073–1076. http://dx.doi.org/10.1645/GE-3202RN.

31. Le Bailly M, Leuzinger U, Schlichtherle H, Bouchet F. 2005. *Diphyllobothrium*: neolithic parasite? *J Parasitol* **91**:957–959. http://dx.doi.org/10.1645/GE-3456RN.1.

32. Goncalves ML, da Silva VL, de Andrade CM, Reinhard K, da Rocha GC, Le Bailly M, Bouchet F, Ferreira LF, Araujo A. 2004. Amoebiasis distribution in the past: first steps using an immunoassay technique. *Trans R Soc*

Trop Med Hyg **98**:88–91. http://dx.doi.org/10.1016/S0035-9203(03)00011-7.

33. Poinar HN, Hofreiter M, Spaulding WG, Martin PS, Stankiewicz BA, Bland H, Evershed RP, Possnert G, Pääbo S. 1998. Molecular coproscopy: dung and diet of the extinct ground sloth *Nothrotheriops shastensis*. *Science* **281**:402–406. http://dx.doi.org/10.1126/science.281.5375.402.

34. Ortega YR, Bonavia D. 2003. *Cryptosporidium*, *Giardia*, and *Cyclospora* in ancient Peruvians. *J Parasitol* **89**:635–636. http://dx.doi.org/10.1645/GE-3083RN.

35. Allison MJ, Bergman T, Gerszten E. 1999. Further studies on fecal parasites in antiquity. *Am J Clin Pathol* **112**:605–609. http://dx.doi.org/10.1093/ajcp/112.5.605.

36. da Rocha GC, Harter-Lailheugue S, Le Bailly M, Araújo A, Ferreira LF, da Serra-Freire NM, Bouchet F. 2006. Paleoparasitological remains revealed by seven historic contexts from "Place d'Armes", Namur, Belgium. *Mem Inst Oswaldo Cruz* **101**(Suppl 2):43–52. http://dx.doi.org/10.1590/S0074-02762006001000008.

37. Fernandes A, Ferreira LF, Gonçalves ML, Bouchet F, Klein CH, Iguchi T, Sianto L, Araujo A. 2005. Intestinal parasite analysis in organic sediments collected from a 16th-century Belgian archeological site. *Cad Saude Publica* **21**:329–332. http://dx.doi.org/10.1590/S0102-311X2005000100037.

38. Shin DH, Chai JY, Park EA, Lee W, Lee H, Lee JS, Choi YM, Koh BJ, Park JB, Oh CS, Bok GD, Kim WL, Lee E, Lee EJ, Seo M. 2009. Finding ancient parasite larvae in a sample from a male living in late 17th century Korea. *J Parasitol* **95**:768–771. http://dx.doi.org/10.1645/GE-1763.1.

39. Gonçalves ML, Araújo A, Duarte R, da Silva JP, Reinhard K, Bouchet F, Ferreira LF. 2002. Detection of *Giardia duodenalis* antigen in coprolites using a commercially available enzyme-linked immunosorbent assay. *Trans R Soc Trop Med Hyg* **96**:640–643. http://dx.doi.org/10.1016/S0035-9203(02)90337-8.

40. Le Bailly M, Gonçalves ML, Harter-Lailheugue S, Prodéo F, Araujo A, Bouchet F. 2008. New finding of *Giardia intestinalis* (Eukaryote, Metamonad) in Old World archaeological site using immunofluorescence and enzyme-linked immunosorbent assays. *Mem Inst Oswaldo Cruz* **103**:298–300. http://dx.doi.org/10.1590/S0074-02762008005000018.

41. Le Bailly M, Bouchet F. 2006. Paléoparasitologie et immunologie: l'exemple d'*Entamoeba histolytica*. *Archeosciences* **30**:129–135. http://dx.doi.org/10.4000/archeosciences.281.

42. **Mitchell PD, Stern E, Tepper Y.** 2008. Dysentery in the crusader kingdom of Jerusalem: an ELISA analysis of two medieval latrines in the City of Acre (Israel). *J Archaeol Sci* **35:**1849–1853. http://dx.doi.org/10.1016/j.jas.2007.11.017.

43. **Cooper A, Poinar HN.** 2000. Ancient DNA: do it right or not at all. *Science* **289:**1139. http://dx.doi.org/10.1126/science.289.5482.1139b.

44. **Hofreiter M, Serre D, Poinar HN, Kuch M, Pääbo S.** 2001. Ancient DNA. *Nat Rev Genet* **2:**353–359. http://dx.doi.org/10.1038/35072071.

45. **Pääbo S, Poinar H, Serre D, Jaenicke-Despres V, Hebler J, Rohland N, Kuch M, Krause J, Vigilant L, Hofreiter M.** 2004. Genetic analyses from ancient DNA. *Annu Rev Genet* **38:**645–679. http://dx.doi.org/10.1146/annurev.genet.37.110801.143214.

46. **Drancourt M, Raoult D.** 2005. Palaeomicrobiology: current issues and perspectives. *Nat Rev Microbiol* **3:**23–35. http://dx.doi.org/10.1038/nrmicro1063.

47. **Willerslev E, Cooper A.** 2005. Ancient DNA. *Proc Biol Sci* **272:**3–16. http://dx.doi.org/10.1098/rspb.2004.2813.

48. **Poinar HN.** 2002. The genetic secrets some fossils hold. *Acc Chem Res* **35:**676–684. http://dx.doi.org/10.1021/ar000207x.

49. **Volossiouk T, Robb EJ, Nazar RN.** 1995. Direct DNA extraction for PCR-mediated assays of soil organisms. *Appl Environ Microbiol* **61:**3972–3976.

50. **Rosenstraus M, Wang Z, Chang SY, DeBonville D, Spadoro JP.** 1998. An internal control for routine diagnostic PCR: design, properties, and effect on clinical performance. *J Clin Microbiol* **36:**191–197.

51. **Honoré-Bouakline S, Vincensini JP, Giacuzzo V, Lagrange PH, Herrmann JL.** 2003. Rapid diagnosis of extrapulmonary tuberculosis by PCR: impact of sample preparation and DNA extraction. *J Clin Microbiol* **41:**2323–2329. http://dx.doi.org/10.1128/JCM.41.6.2323-2329.2003.

52. **Tito RY, Belknap SL III, Sobolik KD, Ingraham RC, Cleeland LM, Lewis CM Jr.** 2011. Brief communication: DNA from early Holocene American dog. *Am J Phys Anthropol* **145:**653–657. http://dx.doi.org/10.1002/ajpa.21526.

53. **Iñiguez AM, Araújo A, Ferreira LF, Vicente AC.** 2003. Analysis of ancient DNA from coprolites: a perspective with random amplified polymorphic DNA-polymerase chain reaction approach. *Mem Inst Oswaldo Cruz* **98** (Suppl 1):63–65. http://dx.doi.org/10.1590/S0074-02762003000900012.

54. **Iñiguez AM, Reinhard KJ, Araújo A, Ferreira LF, Vicente AC.** 2003a. *Enterobius vermicularis:* ancient DNA from North and South American human coprolites. *Mem Inst Oswaldo Cruz* **98**(Suppl 1):67–69. http://dx.doi.org/10.1590/S0074-02762003000900013.

55. **Luciani S, Fornaciari G, Rickards O, Labarga CM, Rollo F.** 2006. Molecular characterization of a pre-Columbian mummy and in situ coprolite. *Am J Phys Anthropol* **129:**620–629. http://dx.doi.org/10.1002/ajpa.20314.

56. **Cleeland LM, Reichard MV, Tito RY, Reinhard KJ, Lewis CM Jr.** 2013. Clarifying prehistoric parasitism from a complementary morphological and molecular approach. *J Archaeol Sci* **40:**3060–3066. http://dx.doi.org/10.1016/j.jas.2013.03.010.

57. **Willerslev E, Hansen AJ, Binladen J, Brand TB, Gilbert MT, Shapiro B, Bunce M, Wiuf C, Gilichinsky DA, Cooper A.** 2003. Diverse plant and animal genetic records from Holocene and Pleistocene sediments. *Science* **300:**791–795. http://dx.doi.org/10.1126/science.1084114.

58. **Raoult D, Aboudharam G, Crubézy E, Larrouy G, Ludes B, Drancourt M.** 2000. Molecular identification by "suicide PCR" of *Yersinia pestis* as the agent of medieval black death. *Proc Natl Acad Sci U S A* **97:**12800–12803. http://dx.doi.org/10.1073/pnas.220225197.

59. **Gijón Botella H, Afonso Vargas JA, Arnay de la Rosa M, Leles D, González Reimers E, Vicente AC, Iñiguez AM.** 2010. Paleoparasitologic, paleogenetic and paleobotanic analysis of XVIII century coprolites from the church La Concepción in Santa Cruz de Tenerife, Canary Islands, Spain. *Mem Inst Oswaldo Cruz* **105:**1054–1056. http://dx.doi.org/10.1590/S0074-02762010000800017.

60. **Leles D, Araújo A, Ferreira LF, Vicente AC, Iñiguez AM.** 2008. Molecular paleoparasitological diagnosis of *Ascaris* sp. from coprolites: new scenery of ascariasis in pre-Columbian South America times. *Mem Inst Oswaldo Cruz* **103:**106–108. http://dx.doi.org/10.1590/S0074-02762008005000004.

61. **Ubaldi M, Luciani S, Marota I, Fornaciari G, Cano RJ, Rollo F.** 1998. Sequence analysis of bacterial DNA in the colon of an Andean mummy. *Am J Phys Anthropol* **107:**285–295. http://dx.doi.org/10.1002/(SICI)1096-8644(199811)107:3<285::AID-AJPA5>3.0.CO;2-U.

62. **Cano RJ, Tiefenbrunner F, Ubaldi M, Del Cueto C, Luciani S, Cox T, Orkand P, Künzel KH, Rollo F.** 2000. Sequence analysis of bacterial DNA in the colon and stomach of the Tyrolean Iceman. *Am J Phys Anthropol* **112:**297–309. http://dx.doi.org/10.1002/1096-8644(200007)112:3<297::AID-AJPA2>3.0.CO;2-0.

63. Leles D, Araújo A, Ferreira LF, Vicente AC, Iñiguez AM. 2008. Molecular paleoparasitological diagnosis of *Ascaris* sp. from coprolites: new scenery of ascariasis in pre-Colombian South America times. *Mem Inst Oswaldo Cruz* **103:**106–108. http://dx.doi.org/10.1590/S0074-02762008005000004.

64. Kuch M, Poinar H. 2012. Extraction of DNA from paleofeces. *Methods Mol Biol* **840:**37–42. http://dx.doi.org/10.1007/978-1-61779-516-9_5.

65. Kiatpathomchai W, Boonsaeng V, Tassanakajon A, Wongteerasupaya C, Jitrapakdee S, Panyim S. 2001. A non-stop, single-tube, semi-nested PCR technique for grading the severity of white spot syndrome virus infections in *Penaeus monodon*. *Dis Aquat Organ* **47:**235–239. http://dx.doi.org/10.3354/dao047235.

66. Zeaiter Z, Fournier PE, Greub G, Raoult D. 2003. Diagnosis of *Bartonella* endocarditis by a real-time nested PCR assay using serum. *J Clin Microbiol* **41:**919–925. http://dx.doi.org/10.1128/JCM.41.3.919-925.2003.

67. Neumaier M, Braun A, Wagener C, International Federation of Clinical Chemistry Scientific Division Committee on Molecular Biology Techniques. 1998. Fundamentals of quality assessment of molecular amplification methods in clinical diagnostics. *Clin Chem* **44:**12–26.

68. Gijón Botella H, Afonso Vargas JA, Arnay de la Rosa M, Leles D, González Reimers E, Vicente AC, Iñiguez AM. 2010. Paleoparasitologic, paleogenetic and paleobotanic analysis of XVIII century coprolites from the church La Concepción in Santa Cruz de Tenerife, Canary Islands, Spain. *Mem Inst Oswaldo Cruz* **105:**1054–1056. http://dx.doi.org/10.1590/S0074-02762010000800017.

69. Kunin V, Copeland A, Lapidus A, Mavromatis K, Hugenholtz P. 2008. A bioinformatician's guide to metagenomics. *Microbiol Mol Biol Rev* **72:**557–578. http://dx.doi.org/10.1128/MMBR.00009-08.

70. Petrosino JF, Highlander S, Luna RA, Gibbs RA, Versalovic J. 2009. Metagenomic pyrosequencing and microbial identification. *Clin Chem* **55:**856–866. http://dx.doi.org/10.1373/clinchem.2008.107565.

71. Kuczynski J, Lauber CL, Walters WA, Parfrey LW, Clemente JC, Gevers D, Knight R. 2011. Experimental and analytical tools for studying the human microbiome. *Nat Rev Genet* **13:**47–58. http://dx.doi.org/10.1038/nrg3129.

72. Wang Y, Qian PY. 2009. Conservative fragments in bacterial 16S rRNA genes and primer design for 16S ribosomal DNA amplicons in metagenomic studies. *PLoS One* **4:**e7401. http://dx.doi.org/10.1371/journal.pone.0007401.

73. Cano RJ, Rivera-Perez J, Toranzos GA, Santiago-Rodriguez TM, Narganes-Storde YM, Chanlatte-Baik L, García-Roldán E, Bunkley-Williams L, Massey SE. 2014. Paleomicrobiology: revealing fecal microbiomes of ancient indigenous cultures. *PLoS One* **9:**e106833. http://dx.doi.org/10.1371/journal.pone.0106833.

74. Sutton MQ, Reinhard KJ. 1995. Cluster analysis of the coprolites from antelope house: implications for anasazi diet and cuisine. *J Archaeol Sci* **22:**741–750. http://dx.doi.org/10.1016/0305-4403(95)90004-7.

75. Shin DH, Oh CS, Chai JY, Lee HJ, Seo M. 2011. *Enterobius vermicularis* eggs discovered in coprolites from a medieval Korean mummy. *Korean J Parasitol* **49:**323–326. http://dx.doi.org/10.3347/kjp.2011.49.3.323.

76. Haas J, Creamer W, Huamán Mesía L, Goldstein D, Reinhard K, Rodríguez CV. 2013. Evidence for maize (Zea mays) in the Late Archaic (3000-1800 B.C.) in the Norte Chico region of Peru. *Proc Natl Acad Sci U S A* **110:**4945–4949. http://dx.doi.org/10.1073/pnas.1219425110.

77. Bouchet F. 1995. Recovery of helminth eggs from archeological excavations of the Grand Louvre (Paris, France). *J Parasitol* **81:**785–787. http://dx.doi.org/10.2307/3283976.

78. Jiménez FA, Gardner SL, Araújo A, Fugassa M, Brooks RH, Racz E, Reinhard KJ. 2012. Zoonotic and human parasites of inhabitants of Cueva de los Muertos Chiquitos, Rio Zape Valley, Durango, Mexico. *J Parasitol* **98:**304–309. http://dx.doi.org/10.1645/GE-2915.1.

79. De Miranda Chaves S, Reinhard K. 2006. Critical analysis of coprolite evidence of medicinal plant use, Piauí, Brazil. *Palaeogeogr Palaeoclimatol Palaeoecol* **237:**110–118. http://dx.doi.org/10.1016/j.palaeo.2005.11.031.

80. Taglioretti V, Fugassa MH, Beltrame MO, Sardella NH. 2014. Biometric identification of capillariid eggs from archaeological sites in Patagonia. *J Helminthol* **88:**196–202. http://dx.doi.org/10.1017/S0022149X13000035.

81. Horne PD, Tuck JA. 1996. Archaeoparasitology at a 17th century colonial site in Newfoundland. *J Parasitol* **82:**512–515. http://dx.doi.org/10.2307/3284098.

82. Sianto L, Teixeira-Santos I, Chame M, Chaves SM, Souza SM, Ferreira LF, Reinhard K, Araujo A. 2012. Eating lizards: a millenary habit evidenced by Paleoparasitology. *BMC Res Notes* **5:**586. http://dx.doi.org/10.1186/1756-0500-5-586.

83. **Bouchet F.** 1997. Intestinal capillariasis in neolithic inhabitants of Chalain (Jura, France). *Lancet* **349:**256. http://dx.doi.org/10.1016/S0140-6736(05)64868-4.

84. **Bouchet F, West D, Lefèvre C, Corbett D.** 2001. Identification of parasitoses in a child burial from Adak Island (Central Aleutian Islands, Alaska). *C R Acad Sci III* **324:**123–127. http://dx.doi.org/10.1016/S0764-4469(00)01287-7.

85. **Horne PD.** 2002. First evidence of enterobiasis in ancient Egypt. *J Parasitol* **88:**1019–1021. http://dx.doi.org/10.1645/0022-3395(2002)088[1019:FEOEIA]2.0.CO;2.

86. **de Candanedo Guerra RM, Gazêta GS, Amorim M, Duarte AN, Serra-Freire NM.** 2003. Ecological analysis of acari recovered from coprolites from archaeological site of northeast Brazil. *Mem Inst Oswaldo Cruz* **98**(Suppl 1)**:**181–190. http://dx.doi.org/10.1590/S0074-02762003000900027.

87. **Reinhard K, Urban O.** 2003. Diagnosing ancient diphyllobothriasis from Chinchorro mummies. *Mem Inst Oswaldo Cruz* **98**(Suppl 1)**:**191–193. http://dx.doi.org/10.1590/S0074-02762003000900028.

88. **Santoro C, Vinton SD, Reinhard KJ.** 2003. Inca expansion and parasitism in the lluta valley: preliminary data. *Mem Inst Oswaldo Cruz* **98**(Suppl 1)**:**161–163. http://dx.doi.org/10.1590/S0074-02762003000900024.

89. **Dittmar K, Steyn M.** 2004. Paleoparasitological analysis of coprolites from K2, an Iron Age archaeological site in South Africa: the first finding of *Dicrocoelium* sp. eggs. *J Parasitol* **90:**171–173. http://dx.doi.org/10.1645/GE-3224RN.

90. **Moore JG, Grundmann AW, Hall HJ, Fry GF.** 1974. Human fluke infection in Glen Canyon at AD 1250. *Am J Phys Anthropol* **41:**115–117. http://dx.doi.org/10.1002/ajpa.1330410115.

91. **Sianto L, Reinhard KJ, Chame M, Chaves S, Mendonça S, Gonçalves ML, Fernandes A, Ferreira LF, Araújo A.** 2005. The finding of Echinostoma (Trematoda: Digenea) and hookworm eggs in coprolites collected from a Brazilian mummified body dated 600-1,200 years before present. *J Parasitol* **91:**972–975. http://dx.doi.org/10.1645/GE-3445RN.1.

92. **Dean G.** 2006. The science of coprolite analysis: the view from Hinds cave. *Palaeogeogr Palaeoclimatol Palaeoecol* **237:**67–79. http://dx.doi.org/10.1016/j.palaeo.2005.11.029.

93. **Reinhard KJ, Ambler JR, Szuter CR.** 2007. Hunter-gatherer use of small animal food resources: coprolite evidence. *Int J Osteoarchaeol* **17:**416–428. http://dx.doi.org/10.1002/oa.883.

94. **Fugassa MH, Taglioretti V, Gonçalves ML, Araújo A, Sardella NH, Denegri GM.** 2008. *Capillaria* spp. eggs in Patagonian archaeological sites: statistical analysis of morphometric data. *Mem Inst Oswaldo Cruz* **103:**104–105. http://dx.doi.org/10.1590/S0074-02762008000100016.

95. **Johnson KL, Reinhardt KJ, Sianto L, Araújo A, Gardner SL, Janovy J Jr.** 2008. A tick from a prehistoric Arizona coprolite. *J Parasitol* **94:**296–298. http://dx.doi.org/10.1645/GE-1059.1.

96. **Riley T.** 2008. Diet and seasonality in the Lower Pecos: evaluating coprolite data sets with cluster analysis. *J Archaeol Sci* **35:**2726–2741. http://dx.doi.org/10.1016/j.jas.2008.04.022.

97. **Shin DH, Lim DS, Choi KJ, Oh CS, Kim MJ, Lee IS, Kim SB, Shin JE, Bok GD, Chai JY, Seo M.** 2009. Scanning electron microscope study of ancient parasite eggs recovered from Korean mummies of the Joseon Dynasty. *J Parasitol* **95:**137–145. http://dx.doi.org/10.1645/GE-1588.1.

98. **Le Bailly M, Mouze S, da Rocha GC, Heim JL, Lichtenberg R, Dunand F, Bouchet F.** 2010. Identification of *Taenia* sp. in a mummy from a Christian Necropolis in El-Deir, Oasis of Kharga, ancient Egypt. *J Parasitol* **96:**213–215. http://dx.doi.org/10.1645/GE-2076.1.

99. **Fugassa MH, Reinhard KJ, Johnson KL, Gardner SL, Vieira M, Araújo A.** 2011. Parasitism of prehistoric humans and companion animals from Antelope Cave, Mojave County, northwest Arizona. *J Parasitol* **97:**862–867. http://dx.doi.org/10.1645/GE-2459.1.

100. **Tito RY, Macmil S, Wiley G, Najar F, Cleeland L, Qu C, Wang P, Romagne F, Leonard S, Ruiz AJ, Reinhard K, Roe BA, Lewis CM Jr.** 2008. Phylotyping and functional analysis of two ancient human microbiomes. *PLoS One* **3:**e3703. http://dx.doi.org/10.1371/journal.pone.0003703.

101. **Sobolik KD, Gremillion KJ, Whitten PL, Watson PJ.** 1996. Sex determination of prehistoric human paleofeces. *Am J Phys Anthropol* **101:**283–290. http://dx.doi.org/10.1002/(SICI)1096-8644(199610)101:2<283::AID-AJPA10>3.0.CO;2-W.

102. **Lin DS, Connor WE.** 2001. Fecal steroids of the coprolite of a Greenland Eskimo mummy, AD 1475: a clue to dietary sterol intake. *Am J Clin Nutr* **74:**44–49.

103. **Marlar RA, Leonard BL, Billman BR, Lambert PM, Marlar JE.** 2000. Biochemical evidence of cannibalism at a prehistoric Puebloan site in southwestern Colorado. *Nature* **407:**74–78. http://dx.doi.org/10.1038/35024064.

104. **Shillito LM, Almond MJ, Wicks K, Marshall LJ, Matthews W.** 2009. The use of FT-IR as a screening technique for organic residue analysis of archaeological samples. *Spectrochim Acta A Mol Biomol Spectrosc* **72:**120–125. http://dx.doi.org/10.1016/j.saa.2008.08.016.

105. **Vinton SD, Perry L, Reinhard KJ, Santoro CM, Teixeira-Santos I.** 2009. Impact of empire expansion on household diet: the Inka in Northern Chile's Atacama Desert. *PLoS One* **26:**e8069. doi:10.1371/journal.pone.0008069.

106. **Dridi B, Henry M, El Khéchine A, Raoult D, Drancourt M.** 2009. High prevalence of *Methanobrevibacter smithii* and *Methanosphaera stadtmanae* detected in the human gut using an improved DNA detection protocol. *PLoS One* **4:**e7063. http://dx.doi.org/10.1371/journal.pone.0007063.

107. **El Khéchine A, Henry M, Raoult D, Drancourt M.** 2009. Detection of *Mycobacterium tuberculosis* complex organisms in the stools of patients with pulmonary tuberculosis. *Microbiology* **155:**2384–2389. http://dx.doi.org/10.1099/mic.0.026484-0.

108. **Jirků M, Pomajbíková K, Petrželková KJ, Hůzová Z, Modrý D, Lukeš J.** 2012. Detection of *Plasmodium* spp. in human feces. *Emerg Infect Dis* **18:**634–636. http://dx.doi.org/10.3201/eid1804.110984.

109. **Keita AK, Socolovschi C, Ahuka-Mundeke S, Ratmanov P, Butel C, Ayouba A, Inogwabini BI, Muyembe-Tamfum JJ, Mpoudi-Ngole E, Delaporte E, Peeters M, Fenollar F, Raoult D.** 2013. Molecular evidence for the presence of *Rickettsia felis* in the feces of wild-living African apes. *PLoS One* **8:**e54679. http://dx.doi.org/10.1371/journal.pone.0054679.

110. **Demeler J, Ramünke S, Wolken S, Ianiello D, Rinaldi L, Gahutu JB, Cringoli G, von Samson-Himmelstjerna G, Krücken J.** 2013. Discrimination of gastrointestinal nematode eggs from crude fecal egg preparations by inhibitor-resistant conventional and real-time PCR. *PLoS One* **8:**e61285. http://dx.doi.org/10.1371/journal.pone.0061285.

111. **Anastasiou E, Lorentz KO, Stein GJ, Mitchell PD.** 2014. Prehistoric schistosomiasis parasite found in the Middle East. *Lancet Infect Dis* **14:**553–554. http://dx.doi.org/10.1016/S1473-3099(14)70794-7.

112. **Guhl F, Jaramillo C, Vallejo GA, Yockteng R, Cárdenas-Arroyo F, Fornaciari G, Arriaza B, Aufderheide AC.** 1999. Isolation of *Trypanosoma cruzi* DNA in 4,000-year-old mummified human tissue from northern Chile. *Am J Phys Anthropol* **108:**401–407. http://dx.doi.org/10.1002/(SICI)1096-8644(199904)108:4<401::AID-AJPA2>3.0.CO;2-P.

113. **Rollo F, Ubaldi M, Ermini L, Marota I.** 2002. Otzi's last meals: DNA analysis of the intestinal content of the Neolithic glacier mummy from the Alps. *Proc Natl Acad Sci U S A* **99:**12594–12599. http://dx.doi.org/10.1073/pnas.192184599.

Ancient Resistome

8

ABIOLA OLUMUYIWA OLAITAN[1] and JEAN-MARC ROLAIN[1]

It is generally now known that the emergence of antibiotic resistance in bacteria is an ancient biological event, meaning that the existence of resistance genes predates our present-day use of antibiotics. In other words, antibiotic resistance is not a modern phenomenon. These findings have been made possible by emerging fields, such as paleomicrobiology, and paleo-environmental studies of ancient biological samples, deep sub-surface environments, and isolated pristine environments that are free of contemporary sources (contemporary antibiotic compounds or genes) (1). Paleomicrobiology has been employed to gain insight into ancient infections, such as the Black Death and Justinian's plague, and has shown evidence of the involvement of bacteria in these pandemics (2, 3).

The investigation of antibiotic resistance from the perspective of paleobiology is a very recent emerging research area, and very few studies have been performed to detect evidence or the presence of antibiotic-resistant bacteria and resistance genes from ancient samples. Therefore, antibiotic resistance paleomicrobiology is still very much in its infancy.

One of the ways in which to study ancient antibiotic resistance is by sampling and analyzing ancient DNA (aDNA) recovered from archaeological

[1]Aix Marseille Université, Unité de Recherche sur les Maladies Infectieuses et Tropicales Emergentes (URMITE), UMR CNRS 7278, IRD 198, INSERM 1095, Faculté de Médecine, Marseille, France.
Paleomicrobiology of Humans
Edited by Michel Drancourt and Didier Raoult
© 2016 American Society for Microbiology, Washington, DC
doi:10.1128/microbiolspec.PoH-0008-2015

or paleobiological samples. One type of these samples is ancient permafrost; this basically consists of soil, rock, and sediment, including organic materials, that have been permanently frozen for at least two consecutive years. Ancient permafrost has remained permanently frozen over millennia, for between 1 and 3 million years in the Arctic region and longer in the Antarctic region (4). Some of the ancient fauna and flora and the microbial biomass are usually well preserved over the years, which makes permafrost a huge reservoir of ancient viable prokaryotes and genetic materials (5). Another sample is ancient dental calculus, which is dental plaque formed from a complex inorganic and organic matrix, including calcified bacterial biofilm (6, 7). These samples have allowed scientists to analyze the presence of antibiotic resistance genes not only from ancient environmental samples but also from our human ancestors, including the pathogens associated with those humans (7).

ANTIBIOTIC RESISTANCE IN ANCIENT ENVIRONMENTAL SAMPLES

By sampling permafrost sediments in Canada dated to approximately 30,000 years (25,300 radiocarbon), D'Costa et al. provided one of the first authenticated supports for the existence of antibiotic resistance genes thousands of years ago (8). The analysis of aDNA sequences led to the detection of resistance genes covering several antibiotic classes, mainly tetracyclines, vancomycin, and β-lactams (Table 1). Several tetracycline resistance-encoding genes (*tetM*) were detected, with most displaying a high degree of similarity with the ribosomal protection proteins from actinomycetes. Fragments of *vanX*, an important component of the vancomycin resistance operon (*vanHAX*), were identified, and a further search led to the amplification of the *vanHAX* operon in the aDNA samples. Surprisingly, the *vanHAX* operon showed

similarities to operons from contemporary environmental and clinical bacteria. This evidence strengthened the hypothesis that vancomycin resistance in enterococci may have originated from glycopeptide-producing bacteria (9), indicating that *Actinomycetes* may be a reservoir of resistance genes, and the evidence suggests the possibility of the transfer of vancomycin resistance to clinically relevant bacteria (10). Additionally, β-lactamase sequences were recovered from the aDNA, and the sequences displayed 53% to 84% amino acid homology with known β-lactamases (8).

It is noteworthy that several studies have been undertaken with Siberia permafrost, with reports of the detection and isolation of antibiotic-resistant bacteria from ancient permafrost samples (11–14). However, the reliability of these findings remains debatable among researchers in the field. The reasons for the contentious nature of these findings, as emphasized by D'Costa et al., include the absence of controls to rule out modern contamination and authenticate the ancient nature of these samples, an insufficient number of biological and experimental replicates, and the nonreproducibility of similar findings by independent laboratories (8). However, the occurrence of variation in analyzing replicate permafrost data from a particular location has recently been highlighted (4). These disagreements are not unexpected in an emerging field such as paleomicrobiology (1); however, recent technological advancements with next-generation sequencing and contamination management will be valuable tools in overcoming some of the challenges in the field (1).

A culture-based sampling of Lechuguilla Cave, located in New Mexico, USA, and isolated for over 4 million years, was recently performed (15). The study revealed that, on average, more than 70% and 65% of the Gram-positive and Gram-negative bacteria, respectively, from the collection were multidrug-resistant (resistant to three to four classes of antibiotics), including resistance

TABLE 1 Antibiotic resistance genes uncovered from ancient samples

Bacteria	Putative antibiotic resistance genes	Antibiotic class	Age	References
Porphyromonas gingivalis	β-Lactamases, *ble*, efflux pump (MATE[a], MFS[a]), *tetR*, *aph*	β-Lactams, glycopeptides (e.g., bleomycin), tetracyclines, aminoglycosides, bicyclomycin	c. 950–1200 CE	7
Capnocytophaga sputigena	β-Lactamase (*csp-1*, extended-spectrum β-lactamase)	β-Lactams	c. 950–1200 CE	7
Clostridium difficile	Efflux pump (ABC[a], MATE)		c. 950–1200 CE	7
Fusobacterium nucleatum	*flo*, *pbp*	Phenicols (e.g., florfenicol), β-lactams	c. 950–1200 CE	7
Leptotrichia buccalis	Efflux pump (MATE), *pbpA*, *tetR*	β-Lactams, tetracyclines	c. 950–1200 CE	7
Neisseria meningitidis	Efflux pump (MATE), *tetR*, *penA*	β-Lactams, tetracyclines	c. 950–1200 CE	7
Porphyromonas gingivalis	Efflux pump (MATE), *rteC*, *ble*, β-lactamase	Tetracyclines, glycopeptides (e.g., bleomycin), β-lactams	c. 950–1200 CE	7
Rothia mucilaginosa	*pbpA*	β-Lactams	c. 950–1200 CE	7
Streptococcus gordonii	Efflux pump (ABC), *bacA*, *mef/mel*, β-lactamase, *pbps*, *tetR*	Cyclic peptides (e.g., bacitracin), macrolides, β-lactams, tetracyclines	c. 950–1200 CE	7
Streptococcus mitis	*tetR*	Tetracyclines	c. 950–1200 CE	7
Streptococcus pneumoniae	*macB*, β-lactamase, *pbps*, *aac(6')*, *tetR*	Macrolides, β-lactams, aminoglycosides, tetracyclines	c. 950–1200 CE	5
Streptococcus pyogenes	*tetR*	Tetracyclines	c. 950–1200 CE	7
Streptococcus sanguinis	Efflux pump (ABC, MFS, MATE), *ble*, *pbps*, β-lactamase	Glycopeptides (e.g., bleomycin), β-lactams	c. 950–1200 CE	7
Tannerella forsythia	β-Lactamase, *tetR*, *uppP* or *bcrC*, efflux pump (ABC, MATE), *pbpC*, *pbpA*	β-Lactams, tetracyclines, cyclic peptides (e.g., bacitracin)	c. 950–1200 CE	7
Treponema denticola	β-Lactamase, *mef/mel*, *tetR*, *pbp*, efflux pump (e.g., MATE), *nim* gene	β-Lactams, macrolides, tetracyclines, nitroimidazoles (5-nitroimidazole)	c. 950–1200 CE	7
Veillonella parvula	*pbp*	β-Lactams	c. 950–1200 CE	7
Streptomyces rimosus	*tetM*	Tetracyclines	30,000 years	8
Staphylococcus aureus, *Enterococcus faecalis*, *Amycolatopsis orientalis*, *Frankia* spp.	*vanHAX*	Glycopeptides	30,000 years	8
Streptomyces cellulosae	β-Lactamases	β-Lactams	30,000 years	8
Streptomyces spp., *Rhodococcus erythropolis*, *Paenibacillus lautus*, *Arthrobacter* spp.,	β-Lactamases	β-Lactams	4 million years	15

(Continued on next page)

TABLE 1 Antibiotic resistance genes uncovered from ancient samples *(Continued)*

Bacteria	Putative antibiotic resistance genes	Antibiotic class	Age	References
Agrobacterium tumefaciens, *Ochrobactrum anthropi* *Streptomyces* spp., *Brachybacterium paraconglomeratum*	*mph* (glycosylation, phosphorylation)	Macrolides (e.g., erythromycin, telithromycin)	4 million years	15
Agrobacterium tumefaciens, *Ochrobactrum anthropi*	*cat* (acetylation)	Phenicols (e.g, chloramphenicol)	4 million years	15
Streptomyces spp., *Paenibacillus lautus*		Lipopeptides (e.g, daptomycin)	4 million years	15
Chryseobacterium spp.[b]	*cat*	Phenicols (e.g, chloramphenicol)	700 years	16

[a]These efflux pumps may be involved in resistance to multiple classes of antibiotics.
[b]Related to that found in *Chryseobacterium* spp.

to old or new but natural antibiotics, such as penicillins, erythromycin, and daptomycin, and to semisynthetic antibiotics, such as telithromycin. Among the Gram-positive bacteria, which included *Streptomyces* spp. and *Rhodococcus erythropolis*, enzymatic inactivation of penicillins and the cephalosporin cephalexin, normally mediated by β-lactamases, was detected. Resistance to macrolide antibiotics was detected among some of the isolates from the cave, which was observed to occur via enzymatic inactivation. Resistance to both erythromycin and its semisynthetic derivative telithromycin was detected among the *Streptomyces* spp. and in *Brachybacterium paraconglomeratum*, with the former inactivating macrolides via glycosylation and the latter via phosphorylation. Furthermore, the inactivation of daptomycin, a lipopeptide antibiotic approved for human use by the U.S. Food and Drug Administration (FDA) in 2003, was observed among the isolated *Streptomyces* strains and *Paenibacillus lautus*; the contemporary *P. lautus* strain that was examined also possessed a similar attribute.

In addition to the various intrinsic antibiotic resistance mechanisms that were likely displayed by the Gram-negative bacteria isolated from the ancient cave, the acetylation of chloramphenicol leading to resistance was identified among the bacte-

ria belonging to the genera *Agrobacterium* and *Ochrobactrum*.

ANTIBIOTIC RESISTANCE IN ANCIENT HUMAN SAMPLES

The study of the remains of ancient humans has further allowed scientists to directly identify pathogens and to detect the presence of antibiotic resistance genes in those pathogens. In a metagenomic study on the ecology of the ancient oral microbiome of four ancient dental calculi from four adult human skeletons dated back to c. 950 to 1200 CE in Dalheim, Germany, Warinner et al. uncovered interesting facts about the presence of antibiotic resistance in ancient pathogens that were associated with dental calculus in humans (7). Of the 40 opportunistic pathogens identified in the ancient dental calculi, 16 possessed resistance genes displaying homology to contemporary putative antibiotic resistance genes, with resistance to several antibiotic classes including beta-lactams, aminoglycosides, macrolides, tetracyclines, and cyclic peptide (e.g., bacitracin) (Table 1). Important genetic features, such as transposons, phages, and plasmids, which drive the proliferation of resistance genes, were also found among the pathogens, with or without resistance genes. This paleomicrobiological study presents the first direct evidence of the antiquity of antibiotic resistance in bacteria related to human health.

Furthermore, a viral metagenomic study aimed at detecting viruses infecting both eukaryotic and prokaryotic cells was recently conducted on ancient petrified fecal material (coprolite) from Namur, Belgium, dated back to about 700 years (14th century) (16). Among the virulence genes detected were a *cat* gene encoding chloramphenicol O-acetyltransferase, known to confer resistance to chloramphenicol antibiotic. The *cat* gene was found to be closely related to that found in *Chryseobacterium sp*. This finding was in addition to numerous bacteriophages detected in the virome that are known to be involved in the diffusion and transfer of antibiotic resistance genes through transduction (17, 18). This finding implies the existence of bacteria harboring antibiotic resistance genes in the human gut as far back as the Middle Ages.

CONCLUSION

It is evident that antimicrobial resistance predates our modern world. However, the advent of antibiotics in human and non-human medicine has created a selective pressure that has led to the proliferation of resistant bacteria in our contemporary environments. Although the study of the antiquity of antibiotic resistance is still in its infancy, it is likely that further studies, including metagenomic studies involving hitherto unexplored ancient archaeological or paleontological samples, will shed further light on the resistomes of ancient bacteria. However, strict protocols must be followed to ensure the validity of the findings.

PERSPECTIVE

We believe that as research in antibiotic resistance paleomicrobiology gradually progresses, it will aid the discovery of novel mechanisms of resistance to antibiotics, and that this research will be used proactively in the development of new antibiotics. Because resistance is ancient, fundamental questions about the other physiological roles that resistance plays in bacteria can be studied further. Finally, studying specific ancestral resistance genes will improve our understanding of the evolution of antibiotic resistance genes.

CITATION

Olaitan AO, Rolain JM. 2016. Ancient resistome. Microbiol Spectrum 4(4):PoH-0008-2015.

REFERENCES

1. **Warinner C, Speller C, Collins MJ.** 2015. A new era in palaeomicrobiology: prospects for ancient dental calculus as a long-term record of the human oral microbiome. *Philos Trans R Soc Lond B Biol Sci* **370:**20130376. doi:10.1098/rstb.2013.0376.
2. **Drancourt M, Roux V, Dang LV, Tran-Hung L, Castex D, Chenal-Francisque V, Ogata H, Fournier PE, Crubezy E, Raoult D.** 2004. Genotyping, Orientalis-like *Yersinia pestis*, and plague pandemics. *Emerg Infect Dis* **10:**1585–1592.
3. **Drancourt M, Raoult D.** 2005. Palaeomicrobiology: current issues and perspectives. *Nat Rev Microbiol* **3:**23–35.
4. **Jansson JK, Tas N.** 2014. The microbial ecology of permafrost. *Nat Rev Microbiol* **12:**414–425.
5. **Gilichinsky D, Vishnivetskaya T, Petrova M, Spirina E, Mamykin V, Rivkina E.** 2008. Bacteria in permafrost, p 83–102. *In* Margesin R, Schinner F, Marx JC, Gerday C (ed), *Psychrophiles: from Biodiversity to Biotechnology.* Springer, Berlin, Germany.
6. **Jin Y, Yip HK.** 2002. Supragingival calculus: formation and control. *Crit Rev Oral Biol Med* **13:**426–441.
7. **Warinner C, Rodrigues JF, Vyas R, Trachsel C, Shved N, Grossmann J, Radini A, Hancock Y, Tito RY, Fiddyment S, Speller C, Hendy J, Charlton S, Luder HU, Salazar-Garcia DC, Eppler E, Seiler R, Hansen LH, Castruita JA, Barkow-Oesterreicher S, Teoh KY, Kelstrup**

CD, Olsen JV, Nanni P, Kawai T, Willerslev E, von MC, Lewis CM Jr, Collins MJ, Gilbert MT, Ruhli F, Cappellini E. 2014. Pathogens and host immunity in the ancient human oral cavity. *Nat Genet* **46:**336–344.

8. D'Costa VM, King CE, Kalan L, Morar M, Sung WW, Schwarz C, Froese D, Zazula G, Calmels F, Debruyne R, Golding GB, Poinar HN, Wright GD. 2011. Antibiotic resistance is ancient. *Nature* **477:**457–461.

9. Marshall CG, Lessard IA, Park I, Wright GD. 1998. Glycopeptide antibiotic resistance genes in glycopeptide-producing organisms. *Antimicrob Agents Chemother* **42:**2215–2220.

10. D'Costa VM, McGrann KM, Hughes DW, Wright GD. 2006. Sampling the antibiotic resistome. *Science* **311:**374–377.

11. Mindlin SZ, Soina VS, Ptrova MA, Gorlenko Z. 2008. Isolation of antibiotic resistance bacterial strains from East Siberia permafrost sediments. *Genetika* **44:**36–44.

12. Petrova M, Gorlenko Z, Mindlin S. 2009. Molecular structure and translocation of a multiple antibiotic resistance region of a *Psychrobacter psychrophilus* permafrost strain. *FEMS Microbiol Lett* **296:**190–197.

13. Petrova MA, Gorlenko Z, Soina VS, Mindlin SZ. 2008. Association of the strA-strB genes with plasmids and transposons in the present-day bacteria and in bacterial strains from permafrost. *Genetika* **44:**1281–1286.

14. Vishnivetskaya TA, Petrova MA, Urbance J, Ponder M, Moyer CL, Gilichinsky DA, Tiedje JM. 2006. Bacterial community in ancient Siberian permafrost as characterized by culture and culture-independent methods. *Astrobiology* **6:**400–414.

15. Bhullar K, Waglechner N, Pawlowski A, Koteva K, Banks ED, Johnston MD, Barton HA, Wright GD. 2012. Antibiotic resistance is prevalent in an isolated cave microbiome. *PLoS One* **7:**e34953. doi:10.1371/journal.pone.0034953.

16. Appelt S, Fancello L, Le BM, Raoult D, Drancourt M, Desnues C. 2014. Viruses in a 14th-century coprolite. *Appl Environ Microbiol* **80:**2648–2655.

17. Rolain JM, Fancello L, Desnues C, Raoult D. 2011. Bacteriophages as vehicles of the resistome in cystic fibrosis. *J Antimicrob Chemother* **66:**2444–2447.

18. Mazaheri Nezhad FR, Barton MD, Heuzenroeder MW. 2011. Bacteriophage-mediated transduction of antibiotic resistance in enterococci. *Lett Appl Microbiol* **52:**559–564.

The History of Epidemic Typhus

9

EMMANOUIL ANGELAKIS,[1] YASSINA BECHAH,[1] and DIDIER RAOULT[1]

Bacteria of the order *Rickettsiales* were first described as short Gram-negative bacillary microorganisms that retained basic fuchsin when stained by the method of Gimenez and grew in association with eukaryotic cells (1). In 1993, the order *Rickettsiales* was divided into three families—namely, *Rickettsiaceae*, *Bartonellaceae*, and *Anaplasmataceae*. Two distinct groups are found within the *Rickettsia* genus, including the spotted fever group and the typhus group. In the typhus group, the two species *Rickettsia typhi* and *Rickettsia prowazekii* are pathogenic in humans (Table 1). *R. typhi* causes murine typhus, which is a flea-transmitted disease that occurs in warm climates (2). *R. prowazekii* is responsible for epidemic typhus, a disease of the cold months, during which heavy clothing and poor sanitary conditions are conducive to lice proliferation (3).

Epidemic typhus is recognized for its high mortality rate throughout human history, particularly before the advent of modern sanitary practices and the availability of antimicrobial drugs (3). Typhus spreads in crowded and unsanitary conditions, and historical epidemics of the disease often followed war, climate extremes, famine, and social upheaval (4). Epidemic typhus remains a threat in the rural highlands of South America, Africa, and

[1]Unité de Recherche sur les Maladies Infectieuses Transmissibles et Emergentes, Aix-Marseille Université, UM63, CNRS 7278, IRD 198, INSERM U1095, Marseille, France.
Paleomicrobiology of Humans
Edited by Michel Drancourt and Didier Raoult
© 2016 American Society for Microbiology, Washington, DC
doi:10.1128/microbiolspec.PoH-0010-2015

TABLE 1 **Some epidemiological features of typhus group** *Rickettsiae*

TG rickettsiae	Disease	Vector	Host	Geographical distribution
R. prowazekii	Epidemic typhus Brill-Zinsser disease	*Pediculus humanus corporis*	Humans	Worldwide
	Sylvatic typhus	Ticks?	ND	Worldwide
		None	Humans	
		Lice, fleas	Flying squirrels	Eastern USA
R. typhi	Murine typhus	Fleas (*Xenopsylla cheopis*, other flea spp.)	Rodents (rats)	Worldwide
		Ctenocephalides felis	Opossums	
		Lice, mites?	ND	

Abbreviation: TG, typhus group; ND, not determined.

Asia. As a result, epidemic typhus could reemerge as a serious infectious disease in areas of the world where social strife and underdeveloped public health programs persist (4). Indeed, in the past 20 years, areas of Russia, Burundi, Algeria, and Andean Peru have experienced typhus outbreaks (1).

In addition, there is a relapse form of epidemic typhus, called Brill-Zinsser disease (5–7). After primary infection and clinical cure, by an unknown process some infected persons retain the bacteria in a latent form for the rest of their lives. In an indeterminable number of individuals and under the circumstances of fading immunity years or decades after the primary infection, these latent bacteria may reactivate and induce a relapsing form of the disease (8, 9). In this review, we discuss the history and the recent microbiological methods applied for the anthropological and historical investigations of epidemic typhus.

EPIDEMIOLOGY

R. prowazekii is transmitted by the body louse (*Pediculus humanus corporis*) (Fig. 1) (10), as was demonstrated by Charles Nicolle at the Pasteur Institute in Tunis in 1909. Lice live in clothing, and their prevalence is increased by cold weather, humidity, poverty, and lack of hygiene. Body lice have a tendency to desert febrile hosts and seek healthy individuals, thus efficiently

spreading disease in human populations. However, lice are extremely host-specific and lack motility, which limits the efficiency of transfer to different hosts. The louse also suffers from the *R. prowazekii* infection, and depending on the number of bacteria in its gut, the louse may be killed within 1 week (11). Death of the louse is due to rupture of the gut-lining epithelium, and the passage of blood into the body cavity causes a red body color, which is why

FIGURE 1 **Human body louse (***Pediculus humanus corporis***). Typical size is 2 to 4 mm.**

typhus has also been called "red louse" disease. Because the louse does not survive the infection and is unable to transmit the infection to its progeny, *R. prowazekii* requires a vertebrate host to maintain its life cycle. It appears that humans with latent *R. prowazekii* infection are the main reservoir enabling bacterial survival and maintenance in nature.

A zoonotic reservoir of *R. prowazekii* has also been reported in the literature. *R. prowazekii* was isolated from African *Hyalomma* ticks (12, 13) and more recently from Mexican *Amblyomma* ticks (14). However, the role of ticks as a reservoir of *R. prowazekii* remains controversial because ticks experimentally infected with *R. prowazekii* die prematurely or produce progeny unable to transmit bacteria when placed in animal hosts. In America, flying squirrels and their ectoparasites naturally infected with *R. prowazekii* have been identified as reservoirs (15). There has been serological evidence of *R. prowazekii* infection in the flying squirrel *Glaucomys volans* since 1963 (16). Sporadic cases of sylvatic epidemic typhus without fatality continue to be reported in the United States (17, 18). Isolates of *R. prowazekii* from flying squirrels multiply readily in human body lice, but flying squirrel lice are extremely host-specific and do not bite humans. Infection may be spread to humans through ectoparasite feces that are aerosolized when flying squirrels groom themselves. The existence of this zoonotic reservoir for *R. prowazekii* infection in the western hemisphere indicates that epidemic typhus cannot be eradicated by human immunization. Interestingly, the first case of Brill-Zinsser disease following primary infection with sylvatic epidemic typhus was recently reported (19).

Transmission of *R. prowazekii* does not occur directly by bites but by contamination of bite sites, conjunctivae, and mucous membranes with the feces or crushed bodies of infected lice. This is thought to be the main route of infection for health workers attending patients (20). Infection through aerosols of feces-infected dust has been reported and is likely the main route of typhus contraction for physicians (21). Indeed, *R prowazekii* can remain viable for 100 days in lice feces (22). The survival mechanism of bacteria in the feces of lice is not known, although the presence of several *spo*T genes in the genome of *R. prowazekii* may play a role as the *spo*T gene has been described as an effector of adaptation to stress conditions.

Body lice are also responsible for the transmission of louse-borne relapsing fever (*Borrelia recurrentis*), trench fever, bacillary angiomatosis, endocarditis, chronic bacteremia, and chronic lymphadenopathy (*Bartonella quintana*) (23, 24). Body lice usually feed five times per day; proteins in their saliva provoke an allergic reaction and lead to pruritus and the scratching that facilitates the fecal transmission of these agents (25). Two other bacteria have also been detected in body lice: *Acinetobacter* spp. and *Serratia marcescens*. However, it is not known if they can be transmitted to humans by louse bites (26).

CLINICAL MANIFESTATIONS

After inoculation, *R. prowazekii* spreads throughout the body via the bloodstream and enters the endothelial cells of capillaries and small blood vessels to produce vasculitis, usually in the skin, heart, central nervous system, skeletal muscle, and kidneys (27). In severe cases, endothelial damage results in permeability changes and the passage of plasma and plasma proteins from the intravascular compartments to the interstitium. As a result, tissue biopsy specimens have revealed perivascular infiltration, mainly by lymphocytes, plasma cells, polymorphonuclear macrophages, and histiocytes, with or without necrosis of the vessel (21).

The incubation period of epidemic typhus in humans is usually 10 to 14 days. The

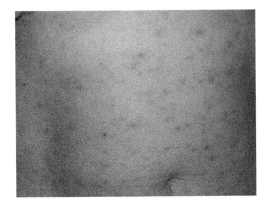

FIGURE 2 **Classic abdominal skin rash of typhus.**

clinical symptoms at the onset of the disease are high fever and headache, and a rash develops on the trunk and limbs. The majority of patients develop malaise and vague symptoms before the abrupt onset of nonspecific constitutional symptoms, including high fever (100%), headaches (100%), and severe myalgias (70% to 100%) (20). In a recent investigation in Burundi, a crouching attitude due to myalgia, named *sutama*, was reported (21). A petechial rash may appear in 20% to 60% of cases (20) (Fig. 2). In Africa, the rash is observed more rarely (20% to 40%) (28). Other possible manifestations are nausea or vomiting (42% to 56%) and coughing/pneumonia (38% to 70%) (20). Most patients with epidemic typhus manifest one or more abnormalities of central nervous system function, such as meningeal irritation and focal or generalized cortical dysfunction including seizures, confusion, drowsiness, coma, and hearing loss (20). Myocarditis, pulmonary involvement, mild thrombocytopenia, jaundice, and abnormal serum liver function test values may occur in severe cases. In uncomplicated epidemic typhus, fever usually resolves after 2 weeks of illness if untreated, but recovery of strength usually takes 2 to 3 months (27). Without treatment, the disease is fatal in 13% to 30% of cases (1).

People who survive epidemic typhus remain infected with *R. prowazekii* for life and under conditions of stress or a waning im-

mune system may experience a recrudescence known as Brill-Zinsser disease. Brill-Zinsser disease is generally milder, but the patients may become the source of a new epidemic if they become infested with body lice (29).

HISTORY OF EPIDEMIC TYPHUS AND DESCRIPTION OF *RIKETTSIA PROWAZEKII*

The historical and geographical origins of typhus are disputed (Fig. 3, Table 2). Some medical historians have suggested epidemic typhus to be an old European disease that caused the Athens plague described by Thucydides (4). The disease may have been identified as early as 1083 in Spain in patients with fever, rash, and parotid tumefaction, although later findings suggest other causes (30). The original description of typhus is thought to have been made in 1546 by

FIGURE 3 **Actual foci of epidemic typhus.**

TABLE 2 **History of epidemic typhus and the description of** *Rickettsia prowazekii*

Year	Event
1489	First account of typhus
1739	First distinctions between typhus and typhoid
1760	Description of exanthematic typhus
1810	Description of diseases
1836	Histological distinction between typhus and typhoid
1909	Role of body lice in typhus and transmission to chimpanzees and monkeys
1910	Description of recrudescent form
1910	Serology test based on *Proteus* for typhus fever
1911	Isolation of *R. prowazekii* in guinea pig
1916	Description of bacteria; named *R. prowazekii*
1916	Weil–Felix test for typhus fever
1930	Typhus vaccine
1934	Tissue cultures
1937	Cell culture (mouse and chick embryos in Koll flasks)
1938	Role of louse feces; culture of bacteria in yolk sacs of embryonated chicken eggs
1948	Treatment with tetracycline and chloramphenicol
1956	Vaccine Madrid E attenuated
1975	Reservoir: flying squirrels (*Glaucomys volans*)
1998	Genome sequence

Fracastoro, a Florentine physician, in his treatise on infectious diseases, *De contagione et contagiosis morbis*. His observations during the Italian outbreaks in 1505 and 1528 allowed him to separate typhus from the other pestilential diseases (4). During the 16th century, typhus fever was gradually distinguished from diseases with similar clinical manifestations as physicians learned to recognize typhus by its sudden onset and characteristic rash (4). However, despite early evidence for typhus in Europe, it is unclear whether typhus was imported from Europe to the New World during colonization or vice versa (31). In 1676, Zavorziz described the disease as it followed armies through Europe, reducing their numbers and spreading into civilian populations through survivors (32). Until the 18th century, epidemic typhus was confused with typhoid fever, and in 1739, Huxham made the first distinction between these diseases. Later, in

1760, Boissier de Sauvages confirmed this observation and introduced the term *exanthematic typhus*. In 1836, William Gerhard of Philadelphia clearly distinguished between typhus and typhoid fevers based on postmortem findings in six patients, which revealed the remarkable absence of ulceration in Peyer's patches, in stark contrast to the observations of typhoid. By the 1860s, physicians had distinguished typhus from another louse-borne disease, relapsing fever (*B. recurrentis*), but diagnostic problems remained. Later, Bacot established that trench fever and typhus are different diseases, the first caused by *B. quintana* and the second by *R. prowazekii*. In 1896, Nathan Brill noted sporadic cases of a typhoid-like illness with negative blood cultures during an epidemic of typhoid fever in New York in the United States (8); this was the first description of so-called Brill-Zinsser disease, the recrudescent form of epidemic typhus. In the 1930s, Hans Zinsser used bacteriological and epidemiological methods to show that the disease described by Brill was an imported form of classic typhus (8). In 1909, Charles Nicolle at the Pasteur Institute in Tunis demonstrated that epidemic typhus is transmitted by the human body louse, and the same year Howard Ricketts in Mexico showed a relationship between the causative agent of epidemic typhus and that of Rocky Mountain spotted fever (33). Nicolle was able to transmit typhus from humans to chimpanzees and then to macaques through blood, and finally from macaque to macaque via a body louse. In 1910, Howard Ricketts contracted typhus and died in Mexico while conducting his experiments. Stanislaus von Prowazek and Henrique da Rocha Lima discovered that typhus is transmitted through the feces of lice rather than through their bites, and in 1914, von Prowazek in turn died of typhus after confirming Ricketts's observations (10). In 1916, da Rocha Lima described the bacterium and named it *R. prowazekii* in honor of Ricketts and Prowazek. In 1922, Wolbach described the human histopathology of *R. prowazekii* infection (34). In 1925, Spencer and Parker developed a vaccine against Rocky Mountain spotted fever, and Rudolph Weigl in Poland, in 1930, developed a similar vaccine against epidemic typhus by grinding up the intestines of lice infected with *R. prowazekii* (10). An attenuated, live *R. prowazekii* vaccine against epidemic typhus is now also available (Madrid E strain), although it is infrequently used. At the same time, the introduction of chemicals such as DDT, which destroy lice and various insects, helped tremendously in the control of typhus (4). In 1938, Cox found that *R. prowazekii* can be grown without difficulty in embryonated chicken eggs (35). This method made it possible to produce very large quantities of live rickettsiae. Finally, after many studies on the fundamental mechanisms of *R. prowazekii* intracellular life and its effects on hosts cells, the genome of an avirulent strain (Madrid E) was entirely sequenced by Andersson et al. in 1998 (36), and the genome of a virulent strain (Rp22) was sequenced in 2010 (37). More recently, a successful genetic transformation of *R. prowazekii* was performed, resulting in a *pld* mutant of the Evir strain (38).

PALEOMICROBIOLOGY

Paleomicrobiology is an emerging field of research in microbiology that is devoted to the detection, identification, and characterization of microorganisms in ancient specimens that can date from centuries to thousands of years in age (39). It was introduced in 1979 (40), and as a discipline, it began in 1993 with the molecular detection of *Mycobacterium tuberculosis* DNA in an ancient human skeleton (39). Paleomicrobiology permits the identification of causative agents of past infectious diseases and of the temporal and geographical distribution of infected groups, and it traces the genetic evolution of microorganisms (39, 41). Therefore, paleomicrobiology opens the way for

the elucidation of controversies concerning various past infections, such as the plague (42–44), epidemic typhus (45), influenza (46), and tuberculosis (47).

Contamination Problems

Most of the data regarding ancient pathogens have been obtained from molecular studies. Some results have been obtained by using microscopy, immunological detection techniques, and the isolation and culture of microorganisms from ancient samples. However, contamination and the authenticity of data regarding ancient pathogens are significant problems in paleomicrobiology. Gilbert et al. found molecular contamination of dental pulp with both previously characterized environmental sequences and unknown sequences when using *pla* and *rpo*B gene-based PCR primers (48). The sources of contamination can be microorganisms from the burial site as well as microorganisms and their DNA from the laboratory and environmental bacteria (39). Several methods can be used to limit the risk for contamination in the laboratory, including absence of a positive control, the introduction of numerous negative controls, and PCR targeting a new specific sequence that has not previously been amplified in the laboratory (49). Moreover, a positive result should be confirmed by the amplification and sequencing of a second specific molecular target and the acquisition of an original sequence that differs from modern homologues (49). Although this approach can preclude contamination, it has poor sensitivity or poor specificity because of the use of nonvalidated target genes, and it is possible that ancient organisms do not differ substantially or at all from their modern homologues.

Materials Used in Paleomicrobiological Studies

The human samples that have been used in paleomicrobiological studies include frozen tissues, mummified corpses, fixed tissues from pathology laboratory collections, and buried skeletons (39). Mummified muscle, skin, brain, and lung have also been used for the protein-based identification of microbes (50). Bone is the most widely used material in the detection of human pathogens, although it can be contaminated by soil flora and the organic content can be washed out (39). Dental pulp was also proposed as an important source in the detection of ancient bacterial or viral DNA or RNA (39). Dental pulp is a well-vascularized soft tissue, and it has been demonstrated that dental pulp is the equivalent of a small blood specimen that is sterile in undiseased mammals with definite dental growth, including humans (49, 51, 52). In addition, dental pulp is better protected from external contamination in teeth without trauma and in unerupted teeth (48). However, the amount of extractable DNA decreased by 90% after only 6 weeks of tooth storage in soil (53).

Assays Used in Paleomicrobiology

Molecular biology techniques are commonly used to detect microbial DNA, but the risk for contamination, chemical modification, and fragmentation of DNA and the presence of PCR inhibitors in ancient samples have led researchers to explore alternative methods based on antigenic protein detection. Indeed, the application of universal 16S rRNA gene-based detection and identification of bacteria in paleomicrobiology has been limited by the contamination of ancient material (48). Moreover, Hoss et al. demonstrated the impact of the environment on DNA conservation, showing inverse correlations between both the average temperature and the humidity levels of archeological sites and DNA retrieval (54). In contrast, the "suicide PCR" amplification procedure ensures the absence of amplicon contamination of ancient material in the laboratory because the PCR primers are designed to hybridize targets outside genomic regions previously targeted in the laboratory (49). The availability of genome-based genotyping systems has facilitated pioneering work in the genotyping of some ancient bacteria (49). Ancient mycobacteria were genotyped by sequencing the phospholipase C *mtp40* gene, an *M. tuberculosis*–specific region (55). The genotyping of *M. tuberculosis* with the use of five molecular targets allowed a study of the distribution of tubercle bacilli among family members in Hungarian mummies (56). Multiple spacer typing (MST), which is based on complete genome analysis, has been developed for the genotyping of ancient *Yersinia pestis* (57). The *R. prowazekii* rpmE/tRNA^{fMet} intergenic spacer sequence has been used to genotype *R. prowazekii* DNA in teeth collected from individuals found in a mass grave in Douai, France (58).

Protein-based methods are considered more suitable for detecting historical samples because proteins are more resistant than DNA to environmental degradation. Microbial antigens in ancient specimens can be detected with several techniques, including immunochromatographic detection, enzyme-linked immunosorbent assay (ELISA), and immunohistochemical analysis (59). ELISA and immunohistochemical analysis were used to identify the F1 antigen of *Y. pestis* in ancient skeletons of plague victims from Venice and Genoa (60). However, ELISA is unsuitable for ancient samples because of the detection limit of the assay and the availability of only small quantities of samples. Immunohistochemistry was used to detect the syphilis bacterium *Treponema pallidum* in the mummy of Maria d'Aragona (1503 to 1568), which was recovered from an abbey in Naples, Italy (61). Immunohistochemistry was also used to detect the Rocky Mountain spotted fever bacterium *Rickettsia rickettsii* in different tissues collected from a patient who died in 1901 with a false diagnosis of typhus, thus providing an example of the retrospective diagnosis of closely related diseases (62). Recently in our laboratory, to increase the sensitivity of protein detection, we used immuno-PCR to detect *Y. pestis* proteins in dental pulp specimens collected from Black Death victims (42). Mass spectrometry has also been proposed for the analysis of ancient microbial proteins. Several studies have reported the use of matrix-assisted laser desorption/ionization tandem time-of-flight mass spectrometry (MALDI-TOF MS) for ancient animal proteins or peptides, paving the way for a unique application in paleomicrobiology (59, 63). Ancient mammals were previously identified by analyzing dried muscle and skin collected from well-preserved museum specimens, hair shafts from permafrost-preserved animals, and the bones and teeth of buried animals (63).

Paleomicrobiological Evidence of Past Epidemic Outbreaks of Typhus

Louse-borne typhus fever is undoubtedly one of the most pestilential diseases of humankind. In general, massive population movement, poor hygiene, and famine favor the development of epidemic typhus outbreaks. Moreover, many large wars and invasions were associated with or followed by possible outbreaks of epidemic typhus.

City of Douai

The city of Douai, France, was besieged from 1710 to 1712 and was successively occupied by the French, then the Dutch; it was retaken by French in 1712 during the War of the Spanish Succession. These events occurred within the framework of a generalized European war in which France and Spain opposed other nations; the battle was conducted on the French side by King Louis XIV (58). Recently, a mass grave dating back to the 18th century was explored in Douai (Fig. 4) (58). In the dental pulp obtained from the teeth of the Douai remains, *R. prowazekii* was detected by suicide PCR in 29% of the individuals tested; this was the first identification of an outbreak of epidemic typhus in the 18th century in the context of a pan-European

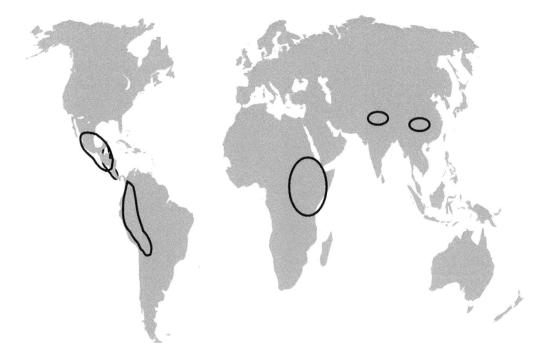

FIGURE 4 **General view of the burial site in Douai, France.**

great war (58). This *R. prowazekii* was genotype B, a genotype previously found in a
Spanish isolate obtained in the first part of
the 20th century (58).

Napoleon Bonaparte's Russian Campaign

Epidemic typhus had long been a scourge of
armies before the 19th century. The most
devastating outbreak followed the invasion of
Russia by Napoleon. Following the retreat of
Napoleon's Grand Army, approximately
70,000 men arrived in Vilnius, Lithuania,
where French troops were garrisoned. Recently, the remains of 717 individuals in
Napoleon's army were detected during
excavations of a mass grave in Vilnius
(Fig. 5) (45). Historical research attributed
this discovery to the retreat of the Grand
Army. Suicide PCR was used to detect louse-
borne pathogens in Napoleon's soldiers, and
four teeth obtained from three soldiers were
found to be positive by suicide PCR with
primers for the *dna*A gene of *R. prowazekii*
(45). In addition, segments of five body lice
that were identified morphologically and by
PCR amplification and sequencing were
also tested, but no *R. prowazekii* DNA was

**FIGURE 5 Imperial-type button found in the grave
in Vilnius.**

detected in these lice (45). This study
revealed that Napoleon's soldiers had epidemic typhus. Although finding bacterial
DNA in teeth does not necessarily mean
that the organism was the cause of death,
when the DNA of a deadly agent such as
R. prowazekii is present, it is very likely that
the organism was the cause of death (45).

Rikettsia. prowazekii Co-infection with Other Bacterial Agents

Diseases transmitted by body lice were
extremely frequent in the past. When sanitary conditions decline, lice are likely to
proliferate quickly, and the population of
lice can increase by 10% per day when
hygienic conditions are poor (64). The threat
posed by body lice is not the louse itself but
the three associated bacterial diseases that
have recently re-emerged: relapsing fever,
caused by *B. recurrentis*; trench fever, caused
by *B. quintana*, and epidemic typhus (22, 65).
In the Douai remains, in which the circulation of louse-borne diseases was suspected,
B. quintana DNA was identified in one
case and *R. prowazekii* DNA in six cases;
B. recurrentis DNA was not recovered (58).
The remains of Napoleon's soldiers in Vilnius
were exposed to body lice containing
B. quintana (45). Moreover, it was found
that 29% of the soldiers had evidence of
infection with either trench fever or epidemic
typhus because *B. quintana* DNA was detected in the dental pulp of the remains of 35
soldiers and *R. prowazekii* was detected in
three other soldiers (45). Recently, following
the outbreak of civil war in 1993 in Burundi,
infection with *B. quintana* and *R. prowazekii*
was diagnosed in refugee camp inhabitants
living under appalling conditions (7).

CONCLUSION

Paleomicrobiology has the potential to contribute to our understanding of the emergence and re-emergence of infectious
diseases. Working together, molecular

biologists, dentists, and anthropologists have explored burials from catastrophes to identify prevailing epidemics in past centuries. The development of new molecular assays and the search for the DNA of infectious agents in dental pulp have improved our knowledge of the history of infectious diseases and past outbreaks. Despite the fact that typhus epidemics have been depicted in historical sources for approximately two millennia, only a few examples of *R. prowazekii* detection from ancient materials have been provided. Epidemic typhus has inexorably changed human history, from the Peloponnesian War to Napoleon's Empire to 20th century Burundi; it continues to affect people and could re-emerge at any moment, even in developed countries.

CITATION

Angelakis E, Bechah Y, Didier R. 2016. The history of epidemic typhus. Microbiol Spectrum 4(4):PoH-0010-2015.

REFERENCES

1. **Raoult D, Roux V.** 1997. Rickettsioses as paradigms of new or emerging infectious diseases. *Clin Microbiol Rev* **10:**694–719.
2. **Angelakis E, Botelho E, Socolovschi C, Sobas CR, Piketty C, Parola P, Raoult D.** 2010. Murine typhus as a cause of fever in travelers from Tunisia and Mediterranean areas. *J Travel Med* **17:**310–315.
3. **Bechah Y, Capo C, Mege JL, Raoult D.** 2008. Epidemic typhus. *Lancet Infect Dis* **8:**417–426.
4. **Raoult D, Woodward T, Dumler JS.** 2004. The history of epidemic typhus. *Infect Dis Clin North Am* **18:**127–140.
5. **Mokrani K, Fournier PE, Dalichaouche M, Tebbal S, Aouati A, Raoult D.** 2004. Re-emerging threat of epidemic typhus in Algeria. *J Clin Microbiol* **42:**3898–3900.
6. **Tarasevich I, Rydkina E, Raoult D.** 1998. Outbreak of epidemic typhus in Russia. *Lancet* **352:**1151.
7. **Raoult D, Ndihokubwayo JB, Tissot-Dupont H, Roux V, Faugere B, Abegbinni R, Birtles RJ.** 1998. Outbreak of epidemic typhus associated with trench fever in Burundi. *Lancet* **352:**353–358.
8. **Lutwick LI.** 2001. Brill-Zinsser disease. *Lancet* **357:**1198–1200.
9. **Stein A, Purgus R, Olmer M, Raoult D.** 1999. Brill-Zinsser disease in France. *Lancet* **353:** 1936.
10. **Gross L.** 1996. How Charles Nicolle of the Pasteur Institute discovered that epidemic typhus is transmitted by lice: reminiscences from my years at the Pasteur Institute in Paris. *Proc Natl Acad Sci U S A* **93:**10539–10540.
11. **Houhamdi L, Fournier PE, Fang R, Lepidi H, Raoult D.** 2002. An experimental model of human body louse infection with *Rickettsia prowazekii*. *J Infect Dis* **186:**1639–1646.
12. **Ormsbee RA, Hoogstraal H, Yousser LB, Hildebrandt P, Atalla W.** 1968. Evidence for extra-human epidemic typhus in the wild animals of Egypt. *J Hyg Epidemiol Microbiol Immunol* **12:**1–6.
13. **Reiss-Gutfreund RJ.** 1966. The isolation of *Rickettsia prowazeki* and *mooseri* from unusual sources. *Am J Trop Med Hyg* **15:**943–949.
14. **Medina-Sanchez A, Bouyer DH, Cantara-Rodriguez V, Mafra C, Zavala-Castro J, Whitworth T, Popov VL, Fernandez-Salas I, Walker DH.** 2005. Detection of a typhus group *Rickettsia* in *Amblyomma* ticks in the state of Nuevo Leon, Mexico. *Ann NY Acad Sci* **1063:**327–332.
15. **Bozeman FM, Masiello SA, Williams MS, Elisberg BL.** 1975. Epidemic typhus rickettsiae isolated from flying squirrels. *Nature* **255:**545–547.
16. **Sonenshine DE, Bozeman FM, Williams MS, Masiello SA, Chadwick DP, Stocks NI, Lauer DM, Elisberg BL.** 1978. Epizootiology of epidemic typhus (*Rickettsia prowazekii*) in flying squirrels. *Am J Trop Med Hyg* **27:**339–349.
17. **Reynolds MG, Krebs JS, Comer JA, Sumner JW, Rushton TC, Lopez CE, Nicholson WL, Rooney JA, Lance-Parker SE, McQuiston JH, Paddock CD, Childs JE.** 2003. Flying squirrel-associated typhus, United States. *Emerg Infect Dis* **9:**1341–1343.
18. **Chapman AS, Swerdlow DL, Dato VM, Anderson AD, Moodie CE, Marriott C, Amman B, Hennessey M, Fox P, Green DB, Pegg E, Nicholson WL, Eremeeva ME, Dasch GA.** 2009. Cluster of sylvatic epidemic typhus cases associated with flying squirrels, 2004-2006. *Emerg Infect Dis* **15:**1005–1011.
19. **McQuiston JH, Knights EB, Demartino PJ, Paparello SF, Nicholson WL, Singleton J, Brown CM, Massung RF, Urbanowski JC.** 2010. Brill-Zinsser disease in a patient following infection with sylvatic epidemic typhus

associated with flying squirrels. *Clin Infect Dis*
51:712–715.

20. **Parola P, Raoult D.** 2006. Tropical rickett-sioses. *Clin Dermatol* **24:**191–200.

21. **Houhamdi L, Raoult D.** 2007. Louse-borne epidemic typhus, p 51–61. *In* Raoult D, Parola P (ed). *Rickettsial diseases.* CRC Press, New York, USA.

22. **Raoult D, Roux V.** 1999. The body louse as a vector of reemerging human diseases. *Clin Infect Dis* **29:**888–911.

23. **Brouqui P.** 2011. Arthropod-borne diseases associated with political and social disorder. *Annu Rev Entomol* **56:**357–374.

24. **Veracx A, Raoult D.** 2012. Biology and genetics of human head and body lice. *Trends Parasitol* **28:**563–571.

25. **Foucault C, Brouqui P, Raoult D.** 2006. *Bartonella quintana* characteristics and clinical management. *Emerg Infect Dis* **12:**217–223.

26. **La Scola B, Fournier PE, Brouqui P, Raoult D.** 2001. Detection and culture of *Bartonella quintana, Serratia marcescens,* and *Acinetobacter* spp. from decontaminated human body lice. *J Clin Microbiol* **39:**1707–1709.

27. **Baxter JD.** 1996. The typhus group. *Clin Dermatol* **14:**271–278.

28. **Perine PL, Chandler BP, Krause DK, McCardle P, Awoke S, Habte-Gabr E, Wisseman CL Jr, McDade JE.** 1992. A clinico-epidemiological study of epidemic typhus in Africa. *Clin Infect Dis* **14:**1149–1158.

29. **Weissmann G.** 2005. Rats, lice, and Zinsser. *Emerg Infect Dis* **11:**492–496.

30. **Andersson JO, Andersson SGE.** 2000. A century of typhus, lice and *Rickettsia. Res Microbiol* **151:**143–150.

31. **Burns JN, Acuna-Soto R, Stahle DW.** 2014. Drought and epidemic typhus, central Mexico, 1655-1918. *Emerg Infect Dis* **20:**442–447.

32. **Hansen W, Freney J.** 1996. Le typhus épidémique, sa transmission et la découverte de l'agent étiologique. *Lyon Pharmaceutique* **47:**130–138.

33. **Ricketts HT, Gomez L.** 1908. Studies on immunity in Rocky Mountain spotted fever. *J Infect Dis* **5:**221–244.

34. **Wolbach SB, Todd JL, Palfrey FW.** 1922. Pathology of typhus in man, p 152–256. *In* Wolbach SB, Todd JL, Palfrey FW (ed), *The Etiology and Pathology of Typhus.* Harvard University Press, Cambridge, MA.

35. **Cox HR.** 1938. Use of yolk sac of developing chick embryo as medium for growing rickettsiae of Rocky mountain spotted fever and typhus group. *Public Health Rep* **53:**2241–2247.

36. **Andersson SGE, Zomorodipour A, Andersson JO, Sicheritz-Pontén T, Alsmark UCM, Podowski RM, Näslund AK, Eriksson AS, Winkler HH, Kurland CG.** 1998. The genome sequence of *Rickettsia prowazekii* and the origin of mitochondria. *Nature* **396:**133–140.

37. **Bechah Y, El KK, Mediannikov O, Leroy Q, Pelletier N, Robert C, Medigue C, Mege JL, Raoult D.** 2010. Genomic, proteomic, and transcriptomic analysis of virulent and avirulent *Rickettsia prowazekii* reveals its adaptive mutation capabilities. *Genome Res* **20:**655–663.

38. **Driskell LO, Yu XJ, Zhang L, Liu Y, Popov VL, Walker DH, Tucker AM, Wood DO.** 2009. Directed mutagenesis of the *Rickettsia prowazekii* pld gene encoding phospholipase D. *Infect Immun* **77:**3244–3248.

39. **Drancourt M, Raoult D.** 2005. Palaeomicrobiology: current issues and perspectives. *Nat Rev Microbiol* **3:**23–35.

40. **Swain FM.** 1969. Paleomicrobiology. *Annu Rev Microbiol* **23:**455–472.

41. **Donoghue HD.** 2011. Insights gained from palaeomicrobiology into ancient and modern tuberculosis. *Clin Microbiol Infect* **17:**821–829.

42. **Malou N, Tran TN, Nappez C, Signoli M, Le FC, Castex D, Drancourt M, Raoult D.** 2012. Immuno-PCR—a new tool for paleomicrobiology: the plague paradigm. *PLoS One* **7:**e31744. doi:10.1371/journal.pone.0031744.

43. **Raoult D, Drancourt M.** 2002. Cause of Black Death. *Lancet Infect Dis* **2:**459.

44. **Drancourt M, Raoult D.** 2002. Molecular insights into the history of plague. *Microbes Infect* **4:**105–109.

45. **Raoult D, Dutour O, Houhamdi L, Jankauskas R, Fournier PE, Ardagna Y, Drancourt M, Signoli M, La VD, Macia Y, Aboudharam G.** 2006. Evidence for louse-transmitted diseases in soldiers of Napoleon's Grand Army in Vilnius. *J Infect Dis* **193:**112–120.

46. **Taubenberger JK.** 2003. Fixed and frozen flu: the 1918 influenza and lessons for the future. *Avian Dis* **47:**789–791.

47. **Zink AR, Grabner W, Reischl U, Wolf H, Nerlich AG.** 2003. Molecular study on human tuberculosis in three geographically distinct and time delineated populations from ancient Egypt. *Epidemiol Infect* **130:**239–249.

48. **Gilbert MT, Cuccui J, White W, Lynnerup N, Titball RW, Cooper A, Prentice MB.** 2004. Absence of *Yersinia pestis*-specific DNA in human teeth from five European excavations of putative plague victims. *Microbiology* **150:**341–354.

49. Raoult D, Aboudharam G, Crubezy E, Larrouy G, Ludes B, Drancourt M. 2000. Molecular identification by "suicide PCR" of *Yersinia pestis* as the agent of Medieval Black Death. *Proc Natl Acad Sci U S A* **97:**12800–12803.

50. Le GM, Philip C, Aboudharam G. 2011. Orthodontic treatment of bilateral geminated maxillary permanent incisors. *Am J Orthod Dentofacial Orthop* **139:**698–703.

51. Drancourt M, Aboudharam G, Signoli M, Dutour O, Raoult D. 1998. Detection of 400-year-old *Yersinia pestis* DNA in human dental pulp: an approach to the diagnosis of ancient septicemia. *Proc Natl Acad Sci U S A* **95:**12637–12640.

52. Potsch L, Meyer U, Rothschild S, Schneider PM, Rittner C. 1992. Application of DNA techniques for identification using human dental pulp as a source of DNA. *Int J Legal Med* **105:**139–143.

53. Pfeiffer H, Huhne J, Seitz B, Brinkmann B. 1999. Influence of soil storage and exposure period on DNA recovery from teeth. *Int J Legal Med* **112:**142–144.

54. Hoss M, Jaruga P, Zastawny TH, Dizdaroglu M, Paabo S. 1996. DNA damage and DNA sequence retrieval from ancient tissues. *Nucleic Acids Res* **24:**1304–1307.

55. Taylor GM, Goyal M, Legge AJ, Shaw RJ, Young D. 1999. Genotypic analysis of *Mycobacterium tuberculosis* from medieval human remains. *Microbiology* **145**(Pt 4):899–904.

56. Fletcher HA, Donoghue HD, Holton J, Pap I, Spigelman M. 2003. Widespread occurrence of *Mycobacterium tuberculosis* DNA from 18th-19th century Hungarians. *Am J Phys Anthropol* **120:**144–152.

57. Drancourt M, Roux V, Dang LV, Tran-Hung L, Castex D, Chenal-Francisque V, Ogata H, Fournier PE, Crubezy E, Raoult D. 2004. Genotyping, Orientalis-like *Yersinia pestis*, and plague pandemics. *Emerg Infect Dis* **10:**1585–1592.

58. Nguyen-Hieu T, Aboudharam G, Signoli M, Rigeade C, Drancourt M, Raoult D. 2010. Evidence of a louse-borne outbreak involving typhus in Douai, 1710-1712 during the war of Spanish succession. *PLoS One* **5:**e15405. doi:10.1371/journal.pone.0015405.

59. Tran TN, Aboudharam G, Raoult D, Drancourt M. 2011. Beyond ancient microbial DNA: nonnucleotidic biomolecules for paleomicrobiology. *Biotechniques* **50:**370–380.

60. Sano T, Smith CL, Cantor CR. 1992. Immuno-PCR: very sensitive antigen detection by means of specific antibody-DNA conjugates. *Science* **258:**120–122.

61. Fornaciari G, Castagna M, Tognetti A, Tornaboni D, Bruno J. 1989. Syphilis in a Renaissance Italian mummy. *Lancet* **2:**614.

62. Dumler JS. 1991. Fatal Rocky Mountain spotted fever in Maryland - 1901. *J Am Med Assoc* **265:**718.

63. Tran TN, Aboudharam G, Gardeisen A, Davoust B, Bocquet-Appel JP, Flaudrops C, Belghazi M, Raoult D, Drancourt M. 2011. Classification of ancient mammal individuals using dental pulp MALDI-TOF MS peptide profiling. *PLoS One* **6:**e17319. doi:10.1371/journal.pone.0017319.

64. Birtles RJ, Canales J, Ventosilla P, Alvarez E, Guerra H, Llanos-Cuentas A, Raoult D, Doshi N, Harrison TG. 1999. Survey of *Bartonella* species infecting intradomicillary animals in the Huayllacallan Valley, Ancash, Peru, a region endemic for human bartonellosis. *Am J Trop Med Hyg* **60:**799–805.

65. Angelakis E, Diatta G, Abdissa A, Trape JF, Mediannikov O, Richet H, Raoult D. 2011. Altitude-dependent *Bartonella quintana* genotype C in head lice, Ethiopia. *Emerg Infect Dis* **17:**2357–2359.

Paleopathology of Human Infections: Old Bones, Antique Books, Ancient and Modern Molecules

10

OLIVIER DUTOUR[1]

Paleopathology is the most recent discipline among the sciences of the past; it studies traces of diseases that can be recognized in animal and human remains from ancient times. This research field is located at the interface of medicine, anthropology, and archaeology. Since its beginning in the late 19th century, paleopathology has developed a specific interest in the origin and evolution of infectious diseases affecting human populations. Since the mid 1990s, thanks to the introduction of PCR into the field, the studies of past human infections have undergone a significant revival, evolving from the analysis of ancient writings and bones to that of ancient molecules, allowing evidence-based diagnosis. The goal of this chapter is to provide some examples of an integrative approach that combines these three sources of data for reconstructing the history of human infections.

HUMAN INFECTIONS IN THE PRESENT TENSE: FROM HOST–PATHOGEN INTERACTIONS TO A "ONE HEALTH" CONCEPT

Human pathogens account for much less than 1% of the global diversity of microbial life (1). A total of 1,415 microbial species belonging to 472 different

[1]Laboratoire d'Anthropologie biologique Paul Broca – École Pratique des Hautes Etudes, PSL Research University Paris, Paris, France.
Paleomicrobiology of Humans
Edited by Michel Drancourt and Didier Raoult
© 2016 American Society for Microbiology, Washington, DC
doi:10.1128/microbiolspec.PoH-0014-2015

genera have been reported to cause diseases in humans (2). They are distributed among bacteria (38% of pathogenic species), fungi (22%), helminths (20%), viruses/prions (15%), and protozoa (5%). Of these species, 13% are known for causing emerging/re-emerging diseases in humans, and 58% come from animal hosts. More than three-fourths of the species responsible for emerging or re-emerging infections are zoonotic pathogens (3). The relation between human populations and their pathogens is illustrated by complex models of multiple-host/multiple-pathogens biology (4) showing the close connection between human infectious diseases and animal infections. Besides determining the influence of environmental conditions on host–pathogen interactions (5) that humans share with other animal hosts, the culture that the human species has developed as a unique adaptation strategy in the biological world has a specific and major influence on the emergence, spread, and evolution of infections (6). The impact of human activities on the environment and ecosystems has progressively become, throughout human history, a driving force for infectious diseases. As early as the end of 18th century, Edward Jenner wrote in the introduction to his study on variolae vaccinae, "The deviation of man from the state in which he was originally placed by nature seems to have proved to him a prolific source of diseases" (7). This "deviation"—that is, in modern wording, the impact of human activities on the environment —then intensified considerably with the industrial revolution and accelerated dramatically during the last decades, leading to a rapid development of emerging diseases. Since the 1970s, it is estimated that new infectious diseases have been discovered at an average rate of one every eight months (8). Multiple causes explain this recent increase: (i) those linked to human activities, such as climate change, anthropization of landscapes and ecosystems (especially through deforestation and urbanization), intensification of farming, monoculture practices, reduction of biodiversity, close contact between various animal species and between humans and animals (domestic or wild), environmental pollution, and the wide use of insecticides and pesticides; (ii) those linked to human populations, dynamics, and social organization, such as demography, migrations, worldwide trade, poverty, health system collapses, conflicts, and wars; and (iii) those resulting from the adaptation of microbes and vectors, such as drug and pesticide resistance and increases in virulence or jump species mainly due to disruption of the ecosystems (6, 8). Considering that infectious diseases at the beginning of the 21st century represent a global threat to public health, as illustrated, among others, by outbreaks of Ebola fever, severe acute respiratory distress syndrome (SARS), and avian influenza, and that only an integrative approach including human health, animal health, and environment management is capable of approaching the complexity of host–pathogen interactions, leading international organizations such as the World Health Organization (WHO), the Food and Agriculture Organization of the United Nations (FAO), and the World Organisation for Animal Health (OIE) decided to join forces for "sharing responsibilities and coordinating global activities to address health risks at the human–animal ecosystems interfaces," leading to the "One Health Concept" (9). Among the research themes contributing to a better understanding of infectious processes, the issue of their time and space profiles and of the evolution of pathogens appears to be one priority research topic, which must be developed in a multidisciplinary framework. The past of infectious diseases is indeed of key importance for acquiring a clearer understanding of their present.

HUMAN INFECTIONS IN THE PAST TENSE: A CHALLENGE FOR PALEOPATHOLOGY

In order to be better understood in its diachronic perspective, human–pathogen relationships need to integrate several fundamental concepts, including those of evolution (10), co-evolution (11), pathocenosis (12), Red Queen theory (13), and epidemiological transition (14). Based upon these fundamental concepts, it appears clearly that humans and microbes have a long history in common and that infectious diseases played a major role in human evolution and history (15).

The study of the past of human infections requires an integrative and multidisciplinary approach, as it overlaps two main domains: (i) the biological and environmental sciences and (ii) the sciences of the past. According to Aidan Cockburn (16), "The importance of infectious diseases in the control of populations demands that more attention be given to the study of available evidence of infections in early man." Different sources of evidence are listed in this seminal paper: ancient writings and iconography (paintings, sculpture, pottery) and ancient human remains (mummies, skeletons, coprolites). The diagnostic capability of paleopathology, which studies ancient human remains for evidence of disease, is limited mainly to diseases with a skeletal expression. When one considers the approximately 200 pathogenic species responsible for emerging or re-emerging infectious diseases, only 10% of them, mainly represented by bacteria and fungi, have a possible skeletal expression. Thus, 90% of these infectious diseases fall outside the scope of skeletal paleopathology. However, this sharp limitation has been overcome in the last two decades. Indeed, the study of the past of human infections can be roughly divided in two major periods: before and after the 1990s. Before the 1990s, the topic was mainly a research field in the domains of history and archaeology, studied by historians of medicine and diseases and by paleopathologists by the recording of ancient writing sources and of osteoarchaeological evidence, which were studied independently or in combination (17). Since the 1990s, new approaches have introduced biology into the field, related to the swift development of molecular techniques, such as PCR, which had previously demonstrated its ability to read fragments of genetic information contained in ancient biological remains (18, 19). Then, a first attempt was made in 1993 on archaeological human bones showing evidence of tuberculosis (20), and few months after on paleopathological cases of leprosy (21). Since that date, many diseases and pathogens have been identified by PCR in ancient human remains: leprosy in 1994, the plague in 1998, and then malaria, trypanosomiasis, and viruses like human T-lymphotropic virus 1 (HTLV-1) and the virus that caused Spanish influenza in 2000 in the frozen bodies of victims of the 1918 epidemic (22).

New methods of PCR (e.g., suicide PCR, nested PCR) are applied in this research to increase the sensitivity of detection while controlling as well as possible the risk for contamination by modern DNA. The number of pathogens identified by these methods has grown rapidly: bacteria, viruses, protozoa, and ectoparasites are screened in their various old substrates. They are dated from prehistorical and historical periods (teeth and bones, humans and animals, natural or artificial mummified tissues, frozen bodies in permafrost, samples fixed in historical medical preparations). More than 20 different species of pathogenic microorganisms were identified in these ancient sources between 1993 and 2008 (23). A new discipline has emerged—paleomicrobiology, which is dedicated to the detection, identification, and characterization of microorganisms in ancient remains (23, 24). Thus, the 90% of infectious diseases that previously fell outside the scope of skeletal paleopathology can be now identified by molecular methods; the study of infectious diseases in the past has increased its scope to molecular analysis in addition to skeletal studies, making retrospective diagnosis pass from analogy to demonstration (25). It is a challenge to paleopathology not to restrict the field to its ancient limitations of skeletal pathology because these limits are technical, not conceptual. Indeed, when one goes back to Ruffer's concept of paleopathology, which he defined in 1913 as the science of diseases whose existence can be demonstrated on the basis of human and animal remains from ancient times (26), one can note that the molecular identification of past human pathogens is nowadays demonstrated on this very basis, even in the absence of skeletal lesions.

It can then be a strategic challenge to paleopathology, as the discipline that studies diseases in the past, to adopt this broadest sense and to show that it is able to include in its field the recent developments in molecular techniques, including next-generation sequencing methods. Some molecular biologists are moving in that direction; Michael Knapp, molecular ecologist at the University of Otago, New Zealand, has suggested that this new approach be called genetic paleopathology (27).

EVOLUTION AND PALEOEPIDEMIOLOGY OF INFECTIOUS DISEASES: A COMPREHENSIVE APPROACH FROM ANCIENT BONES TO ANCIENT MOLECULES WITH THE USE OF ANCIENT BOOKS

Rather than developing separate areas of research on ancient bones and ancient molecules, György Palfi and I in the early 1990s felt that, as paleopathologists, we should maintain a multidisciplinary approach for studying infectious diseases in the past and that paleopathology has a key role to play in this issue (28). To be successful, this approach had to be sufficiently open and attractive and to raise fundamental issues on the past and the present of infectious diseases in order to bring together archaeologists, anthropologists, paleopathologists, historians, microbiologists, infectologists, and epidemiologists and to allow fruitful exchanges among all these specialists. We started in 1993 by organizing a first international meeting on the disputed origin of

syphilis. As a result of the success of the meeting and the rapid publication of its proceedings (29), the ICEPID (International Congresses on Evolution and Paleoepidemiology of Infectious Diseases) research network was born. Other conferences followed, which were organized within the same framework: tuberculosis in 1997 (30), leprosy in 1999 (31), plague in 2001 (32), and recently a new conference on tuberculosis in 2012 (33). I proposed (34) to call this approach "a3B," (ancient bones, ancient books, and ancient biomolecules) because it integrates, from a holistic perspective, methods and data from archaeology, history, and biology.

This approach has multiple mutual benefits because it can (i) provide answers to some unsolved historical questions: What was the nature of Athens plague? Where did syphilis come from? When did leprosy arrive in Europe? (ii) confirm a paleopathological diagnosis (e.g., for minor or atypical skeletal lesions due to some infections); (iii) define the cause of mortality in a given archaeological mass grave (plague or not plague); (iv) identify causative strains and determine new phylogenies of pathogens; (v) explore ancient strains virulence; and (vi) help predict the future of some human pathogens. We present four examples of the interest of this approach to the past of human infections that integrates historical, anthropological, and molecular methods.

From Ancient Books to Ancient Molecules: Examples of Past Plagues

The study of past plagues provides a clear example of the efficiency of molecular answers to unsolved historical questions. Indeed, before the 1990s, the study of past plagues belonged exclusively to the field of historical sciences (old books), whereas it was of limited interest to anthropologists and paleopathologists because of the lack of skeletal lesions due to the cause of death. In 1994, the discovery of a 18th mass grave in Marseilles (35) helped to solve this historical question: Was the Great Plague of Marseilles really due to a plague epidemic (infection due to *Yersinia pestis*) or to another acute epidemic disease? Historical records showed that the pit was located at the gardens of the ancient convent of the "Observatins" Friars. The latter served as a "plague" hospital during the epidemics of 1720 to 1722. The pit can be dated historically from May to June of 1722, which corresponds to the relapse of the plague epidemic of 1720 to 1721. A fragmentary list of the names of the victims stored in the city archives showed the acuteness of this epidemic, as the entrance to the hospital was quickly followed by a fatal issue (35, 36). In 1998, archaeological and anthropological research allowed pioneering work on the molecular identification of *Y. pestis* to be conducted (37). This paper reported the first identification of molecular traces of the plague pathogen in ancient human skeletal remains and the first use of dental pulp for ancient DNA studies. Targeted sequences belonging to two *Y. pestis* genes, the *pla* (plasminogen activator) gene and the *rpoB* (beta subunit of RNA polymerase) gene, were detected in the dental pulp of skeletons in two mass graves, the first dating from the 1722 and the second from the 1590 plague epidemic, both epidemics that are historically documented in Provence. According to these results, the 18th century Great Plague of Marseilles was indeed due to the plague pathogen, *Y. pestis*. When published, the results were controversial (38), but the latest studies have confirmed their validity and the weaknesses of counterarguments based only upon the negative results obtained from other mass graves.

The plague of Athens was another historical question. Its exact cause remained unknown for decades. This epidemic started in Athens in 430 BC and, over 4 years, killed a quarter of the population of the city and its army. Historical records, and especially Thucydides' eyewitness testimony, were interpreted by historians as likely indicating a typhus epidemic. This seemed more prob-

able than a plague or measles epidemic, or typhoid. The latter was one of the most improbable candidates according to the semiology precisely described by Thucydides (39). Surprisingly, Papagrigorakis et al. in 2006 described the molecular signature of a pathogen in dental pulp from the skeletal remains of a mass burial dated from this period (Cemetery of Kerameikos) (40); the presence of the *narG* gene identified *Salmonella enterica* serovar Typhi, which is the pathogen of typhoid fever, indicated that typhoid fever, not typhus, was the cause of the plague of Athens. The molecules contradicted history. From books to molecules, history raises questions and paleomicrobiology provides answers regarding ancient plagues (Table 1).

Moreover, the recent introduction of next-generation sequencing into the field of paleomicrobiology brought additional answers by offering the possibility of a genomic approach to ancient pathogens. For ancient plagues, it has been shown that the genome of the virulence plasmid pPCP1 of the *Y. pestis* strain responsible for the medieval Black Death did not differ from the modern one, thus attesting to virulence similar to that of present plagues (41). Therefore, it is implied that the significant death toll due to the medieval plague cannot be explained by a greater virulence of the strains and that it should be attributed to other causes. The number of deaths during historical pandemics is known to have varied greatly, depending on, besides the virulence of the pathogen, the number of people who become infected, the vulnerability of affected population, and the effectiveness of preventive measures (42). Some of these factors can be approached in past populations by the paleoepidemiological survey (43–46).

The draft genome of the Black Death *Y. pestis* from 1348, recently reconstructed (47), revealed that the causative strain belongs to one that is extinct (phylogenetically located on the beginning of branch 1). Paleogenomic analysis of Justinian's plague (48) revealed that it also was due to an

TABLE 1 Plague epidemics and the first molecular identification of ancient DNA (aDNA) of *Yersinia pestis*

Name	Period	Mortality	Cause (aDNA) and reference
Plague of Athens	430 BC	One-fourth of population	*Salmonella typhi* (40)
Antonine plague	165–180 AD	One-fourth of infected people, up to 5 million	Still unknown
Justinian's plague	541–543 AD	One-half of European population	*Yersinia pestis* (78)
Black Death (medieval)	1348–1354 AD	One-third of European population	*Yersinia pestis* (79)
Subsequent outbreaks (100)	14th–18th century	75 million people	*Yersinia pestis* (37)
Third pandemic (started in China)[a]	19th century	10 million people in India	*Yersinia pestis* (80)

[a]The third pandemic was identified by Alexandre Yersin in 1894 (80).

extinct strain located on branch 0. Therefore, the antique (Justinian) and medieval (Black Death) plague epidemics, representing two major events of the history of Western World, were two independent and successive emergences of distinct *Y. pestis* clones coming from an animal reservoir somewhere in Asia (48). Ancient molecules are powerful allies of history.

From Modern Molecules to Ancient Bones: Human Tuberculosis Predates Animal Domestication

Regarding the history of tuberculosis, significant advances have been made in the last decade by using molecular phylogeny, which has enriched paleopathological research. The previous evolutionary model explained the origin of tuberculosis as a transmission from cattle to humans during the emergence of agriculture, in the Neolithic period. This scenario has been paleopathological dogma for almost a century. For paleopathologists, since the beginning of the 20th century, the lack of evidence constituted proof of absence. As stipulated in *The Cambridge Encyclopaedia of Human Paleopathology*, published in 1998, "No iconography or skeletal lesions consistent with tuberculosis from the Paleolithic period have ever been found" (49).

From 2002, a new scenario appeared in the community of microbiologists working on the molecular phylogeny of infectious agents, using the tools of molecular biology and bioinformatics. These works were initiated by a team at the Pasteur Institute in Paris, which is working on modern human and animal strains of the *Mycobacterium tuberculosis* complex (MTBC) (50). The authors studied the polymorphism of 20 variable regions resulting from insertion–deletion events in the genomes of 100 strains responsible for human and animal tuberculosis infection (represented by five species: *M. tuberculosis, M. africanum, M. canettii, M. microti,* and *M. bovis*). Results showed that animal strains (*microti* or *bovis*) lost more genetic material by successive deletions than human strains (*canettii, tuberculosis*). Therefore, the bacilli affecting cattle and other wild animals are, in their genetic structure, more "evolved" than the strains infecting humans, so that the latter must be placed at a more ancestral position.

In a later work (51), the comparison of six housekeeping genes of different strains of MTBC, including *M. canettii*, showed that the phylogeny of the tubercle bacillus began in Africa about 2.6 to 2.8 million years ago, represented by an ancient bacterial species of smooth type (*Mycobacterium prototuberculosis*). It evolved 35,000 to 40,000 years ago outside Africa by a clonal expansion, corresponding to the emergence of *M. tuberculosis*. MTBC would have evolved after that date, contemporaneously with the arrival in Europe of anatomically modern humans (Cro-Magnon), in two lineages: one strictly human and the other predominantly animal, although occasionally human *M. tuberculosis* then diversified from the Fertile Crescent about at least 10,000 years ago, when population growth occurred during the Neolithic period, and propagated by taking advantage of the increased density of the human population (52). These genetic results confirmed that the emergence of human tuberculosis is much older than was admitted in the classic paleopathological scenario. This radical conceptual change coming from modern molecules has revived paleopathological interest in the search for traces of the earliest tuberculosis infection before to animal domestication.

A possible diagnosis of tuberculous meningitis was raised in 2008 during examination of a fragmentary skull cap of *Homo erectus* found in Turkey and dated from several hundreds of thousands years (53). However, in the absence of molecular confirmation and based on the very tenuous nature of the paleopathological lesions observed, this hypothesis was formally rejected by some authors (54).

Finally, the only indisputable evidence of Paleolithic tuberculosis is not human. It has been identified in the remains of bison dating from 17,000 years BP, discovered in a natural trap cave in Wyoming (55). Biomolecular analysis confirmed infection by a *Mycobacterium* belonging to the MTBC, and its paleogenetic profile suggested that it would be an ancestor of *M. bovis*. This information on the evolution of animal tuberculosis in the New World, however, does not solve the issue of human tuberculosis in the Old World.

The Fertile Crescent is a good model area to test both the hypothesis of the predominant role of settlement in the spread of human infections and the hypothesis of the role of bacterial exchanges between humans and animals. In these areas, settlement preceded the domestication process by about a millennium. There is evidence of sedentary occupation sites in the region of the Euphrates at the preceramic Neolithic period (Pre-Pottery Neolithic A, or PPNA), and primitive agriculture started there at the next period (PPNB).

In the Neolithic settlement of Atlit Yam (near Haifa, now submerged by the sea) dating from 9,000 years BP and belonging to PPNC, a period when agriculture and pastoralism were already well installed, paleopathological lesions suggestive of active tuberculosis were identified in two individuals—a child 12 months old with signs of tubercular meningitis and a woman 25 years old, who is presumed to have been the child's mother. Lipid biomarkers and aDNA confirmed the diagnosis by revealing a trace of *M. tuberculosis* in both individuals (56). Although domestication has been largely proved in this Neolithic village, nothing suggests here that human contamination really came from cattle.

More information was obtained from another site in the Fertile Crescent that was recently studied (57)—Dja'de el Mughara, located near the city of Aleppo in Syria. Dating from early PPNB, between about 11,000 and 10,000 years BP—it predates the early beginning of domestication for its ancient levels. This site has delivered numerous human remains, representing a hundred individuals, among them six skeletons (five children and one adult) with paleopathological lesions suggestive of tuberculosis. All of these cases predate animal domestication.

Four of these cases were studied with combined microcomputed tomography and three-dimensional imaging, biochemical analysis of lipid biomarkers, and molecular research of specific DNA sequences. These methods revealed early signs of spinal infection in a 5-year-old child and specific lipid biomarkers characterizing infection by a strain of MTBC in a younger child (about 1 year old). In this same individual, a specific genetic sequence of MTBC (IS 6110) was identified in multiple copies by using a method of nested PCR. Two fragments of this sequence were amplified and sequenced from two different samples (rib and vertebra).

These multidisciplinary studies clearly demonstrate the existence of human tuberculosis in the Fertile Crescent before the domestication process, about 11,000 years ago, and at its early beginnings at 9,000 years BP. Modern molecules thus helped the paleopathologists to refute one of their dogmas, stimulated new research of old bones, and finally led to confirmation that animal domestication was not the origin of human tuberculosis.

From Old Bones to Ancient Books: Ancient Medical Literature Solved Paleopathological Questions about Skeletal Changes

In 1997, Brenda Baker, from Arizona State University, presented at the ICEPID meeting a paleopathological survey performed on four osteoarchaeological series in which she described a new type of vertebral lesion as follows: "circumferential vertebral lesions consisting of multiple resorptive pits on the anterior aspects of thoracic and lumbar bodies." Based on the co-occurrence of these vertebral changes with other pathological conditions indicating tuberculosis, the author suggested that these changes ("smooth-walled resorptive lesions/severe circumferential pitting"), described for the first time in paleopathological literature (58), might be of tubercular origin and that this pattern, ignored by modern clinicians probably because of its precociousness, likely represented early stages of skeletal tuberculosis that could be observed only by paleopathologists. Further studies showed that among human osteological collections with a known cause of death, a frequent association had been found between these vertebral lesions and tuberculosis, especially in younger age groups (59–61). Haas et al. (62) were the first to use molecular techniques to establish a relationship between superficial vertebral alterations and tuberculosis. Thus, a relationship to tuberculosis was confirmed, but the lack of a description of these early stages of vertebral tuberculosis in modern medical literature remained an open question. Was it, as is the case for the microporosity (pitting) observed on subchondral surfaces in degenerative joint disorders, the privilege of paleopathologists to detect the early stages of diseases by direct observations on bones that medical imaging could not reveal to modern physicians?

When he read the proceedings of the meeting on tuberculosis, the late René Lagier, former professor of rheumatology at the University Hospital of Geneva, Switzerland, remembered that a long time before he had seen this special type of lesion described in an old medical book, dating from the end of the 19th century, in which the condition was not attributed to an early stage of vertebral tuberculosis. This book had been published in 1888 by the French physician Victor Ménard, in which he summarized the courses given by Professor Lannelongue at the Faculty of Medicine in Paris (63). In his book, Ménard pointed out that the term *vertebral tuberculosis* refers not only to the "classic" form known as Pott's disease, in which the typical vertebral collapse can be seen, but also to another type of involvement, superficial "carious" lesions or "superficial vertebral tuberculous osteoperiostitis" (Fig. 1). He distinguished between the two anatomical forms of vertebral tuberculosis (classic Pott's disease and superficial vertebral lesions) by the fact that they might appear separately. He pointed out that superficial vertebral "caries" are frequently associated with visceral lesions; vertebral lesions are characterized by a lack of reparation, in contrast to Pott's disease, and noted that affected individuals frequently die of tuberculosis within a short time. The extent of these superficial lesions, appearing as small pits on the anterior and lateral surfaces of vertebrae, is often considerable (they generally affect from five or six to 12 vertebrae). The denuded surface varies in appearance; sometimes it is smooth and plain, but generally it is rough and irregular, with small, sinuous depressions, covered at the sides by newly formed bone layers, and infiltrated by "fungosity." It is suggested that the infection progresses along the blood vessels entering the vertebra, evidenced by enlargement of the vascular channels. This description has totally disappeared from the modern literature on skeletal tuberculosis; it was last mentioned by Sorrel and Sorrel-Dejerine (64) as a rare but severe, rapidly

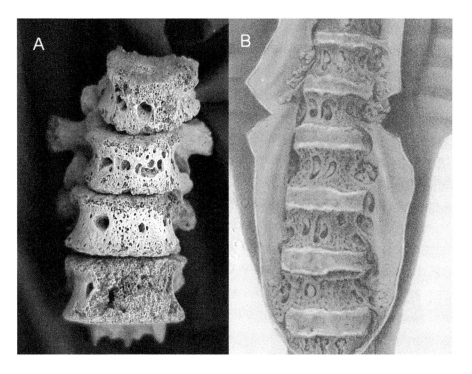

FIGURE 1 Paleopathological (A) and historical (B) illustrations of superficial vertebral lesions due to tuberculosis: (A) Paleopathological case dating from the end of the 18th century (Dutour, 2011). (B) Historical description made by Victor Ménard in 1888 (63).

lethal form of spinal tuberculosis. After that date, it seems that this clinical description totally disappeared from the medical literature. Thus, these lesions, commonly observed in paleopathology, belong to the "forgotten diagnoses" of modern medicine that it is possible to rediscover in old medical treatises. Old pathological bones are better known in ancient medical books than in modern clinical literature.

Such "forgotten diagnoses" should remind us that (i) old clinical descriptions are still of clinical interest, (ii) the natural expression of infectious diseases is strongly influenced by our modern preventive and curative arsenal, and (iii) modern clinical diagnostic criteria are consequently not the most appropriate way to establish retrospective diagnoses of infectious diseases in old bones. Moreover, regarding the increase in antibiotic resistance, which according to the World

Health Organization (65) has become a major threat to public health, it is possible that multidrug resistance will make the future of tuberculosis similar to its past. Therefore, from ancient bones to ancient books, past knowledge of the natural history of infectious diseases is an invaluable source for understanding the possible future of (re)emerging diseases.

From Ancient Bones to Molecules: the Leprosy Case

Skeletal paleopathology plays a key role in finding clues to some ancient infections and demonstrating their antiquity, but skeletal studies cannot be fully informative about the origin and spread of a given human pathogen, and not at all about strain diversity. Starting from dated paleopathological cases, paleomicrobiological methods can provide new data about the origin, spread, and diverse strains diversity of human infections in the past. Leprosy is a fine example of the interest of these exchanges.

The paleopathology of leprosy is based on two sets of lesions (see references 59, 66, and 67, among others). The rhinomaxillary syndrome, characterized by resorption of the nasal spine, widening of the nasal margins, and resorption of the alveolar process of the upper incisors, inducing their loss at advanced stages, is directly due to the bacillus. Acro-osteolysis is characterized by resorption of the tubular bones of the hands and feet, including the "pencil and cup" aspect of joints, dislocation secondary to neurotropic changes, and superinfection. These changes affect the parts of the skeleton that are most subject to taphonomical processes; consequently, the frequency of leprosy in the past may have been underestimated because of taphonomical biases (25).

Regarding the antiquity of this human infection, it seems that the most ancient paleopathological case is located in India and dates from 2000 years BC (68). Two other cases before our era are described in the literature; the first is the Celtic skeleton from Casalecchio di Reno, Bologna, Italy, which is the most ancient evidence of leprosy in Europe, dating from the 4th to the 3rd century BC (69), and the second one is in Roman Egypt, dating from the 2nd century BC (70). Late Antiquity and the early Middle Ages provided more evidence, especially in southern, central, and eastern Europe (Italy, Hungary, Czechia, Croatia, Uzbekistan), whereas after the 10th century AD, numerous cases are reported from northern Europe (Great Britain and Scandinavia). Paleopathological evidence is compatible with the hypothesis of an Asian origin of leprosy during prehistorical times, followed by spread in Europe through Italy during Late Antiquity and the early Middle Ages and further extension to northern Europe during the medieval period. It has been suggested that the introduction of leprosy into Europe was contemporaneous with the return of the troops of Alexander the Great from eastern Asia (71).

Ancient DNA of *Mycobacterium leprae* was first evidenced in 1994 (21). The most ancient case that benefited from molecular analysis has been dated to the 1st century AD (72). In addition to confirming the diagnosis, which can be assisted by the detection of specific lipid biomarkers, known to be more robust than DNA (73), ancient DNA analysis makes possible strain typing, depending on the preservation state of the DNA, and therefore provides phylogeographical information. Indeed, *M. leprae* typing can be done by studying single-nucleotide polymorphism, which differentiates four main genotypic branches (74): branch 1 (Asia), branch 2 (Middle East, East Africa), branch 3 (Europe, Middle East, Americas), and branch 4 (West Africa, South America). Medieval leprosy was first identified as belonging to branch 3 only (genotyping of 11 medieval cases by Monot et al.) (74); however, among five other paleopathological cases from medieval northern Europe, three cases were genotyped as branch 2 subtype F (Middle East) by Schuenemann et al. (75). Recently, two new cases from medieval England were added (76), representing a total of 21 cases of European medieval leprosy showing that 15 paleopathological cases from central and northern Europe are genotype branch 3 and six cases are subgenotype branch 2F, all from northern Europe (Sweden, Denmark, and the United Kingdom). After genotyping leprosy cases dated from the 6th to the 11th century AD from central and eastern Europe and Byzantine Anatolia, Donoghue et al. (77) suggested that the westward migration of people from Central Asia, such as Avar nomadic tribes in the first millennium, may have introduced different *M. leprae* strains from Asia Minor into medieval Europe. Beginning with a diagnosis of leprosy derived from old bones, ancient molecules can solve issues regarding the origin and spread of ancient leprosy epidemics.

CONCLUSION

To summarize, it is clear that an integrative approach that includes human and animal health and environmental management is the only way to understand the infectious threat in our modern world, so as to be able to develop accurate strategies. Moreover, it is well-known that in an evolutionary perspective, the knowledge of past situations can be a powerful tool for understanding the present and for predicting some possible aspects of the future. In order to do so, only an integrative approach bringing together specialists in biomedical and environmental sciences and those in historical and archaeological disciplines that integrate human and animal ancient remains in their context, historical documents, and ancient molecules will help in solving problems about the past conditions of human infections.

CITATION

Dutour O. 2016. Paleopathology of human infections: old bones, antique books, ancient and modern molecules. Microbiol Spectrum 4(4):PoH-0014-2015.

REFERENCES

1. **Editor Nature.** 2011. Editorial: Microbiology by numbers. *Nat Rev Microbiol* **9:**628–628.
2. **Taylor LH, Latham SM, Woolhouse ME.** 2001. Risk factors for human disease emergence. *Philos Trans R Soc Lond B Biol Sci* **356:** 983–989.
3. **Woolhouse MEJ, Taylor LH, Haydon DT.** 2001. Population biology of multihost pathogens. *Science* **292:**1109–1112.
4. **Woolhouse MEJ, Gowtage-Sequeria S.** 2005. Host range and emerging and reemerging Pathogens. *Emerg Infect Dis* **11:**1842–1847.
5. **Dunn RR, Davies TJ, Harris NC, Gavin MC.** 2010. Global drivers of human pathogen richness and prevalence. *Proc Biol Sci* **277:**2587–2595.
6. **McMichael AJ.** 2004. Environmental and social influences on emerging infectious diseases: past, present and future. *Philos Trans R Soc Lond B Biol Sci* **359:**1049–1058.
7. **Jenner E.** 1798. *An Inquiry into the Causes and Effects of the Variolæ Vaccinæ, a Disease Discovered in Some of the Western Counties of England, Particularly Gloucestershire, and Known by the Name of the Cow Pox.* Sampson Low, London, UK.
8. **Ministry of Foreign and European Affairs.** 2011. *French Position on the One Health Concept: for an Integrated Approach to Health in View of the Globalization of Health Risks. Strategic Working Document.*
9. **World Health Organization.** 2010. *The FAO-OIE-WHO Collaboration: Tripartite Concept Note. Sharing Responsibilities and Coordinating Global Activities to Address Health Risks at the Animal-Human-Ecosystems Interfaces.* WHO Press, Geneva, Switzerland.
10. **Darwin C.** 1859. *On the Origin of Species by Means of Natural Selection, or the Preservation of Favoured Races in the Struggle for Life.* John Murray, London, UK.
11. **Ehrlich PR, Raven PH.** 1964. Butterflies and plants: a study in coevolution. *Evolution* **18:**586–608.
12. **Grmek MD.** 1969. Préliminaires d'une étude historique des maladies. *Ann Econ Soc Civil* **XXIV:**1473–1483.
13. **Van Valen L.** 1973. A new evolutionary law. *Evol Theory* **1:**1–30.
14. **Omran AR.** 1971. The epidemiologic transition: a theory of the epidemiology of population change. *Milbank Mem Fund Q* **49:**509–538.
15. **Dobson AP, Carper ER.** 1996. Infectious diseases and human population history. *Bioscience* **46:**115–126.
16. **Cockburn TA.** 1971. Infectious diseases in ancient populations. *Curr Anthropol* **12:**45–62.
17. **Grmek MD.** 1994. *Les maladies à l'aube de la civilisation occidentale.* Payot, Paris, France.
18. **Higuchi R, Bowman B, Freiberger M, Ryder OA, Wilson AC.** 1984. DNA sequences from the quagga, an extinct member of the horse family. *Nature* **312:**282–284.
19. **Paabo S.** 1985. Molecular cloning of ancient Egyptian mummy DNA. *Nature* **314:**644–645.
20. **Spiegelman M, Lemma E.** 1993. The use of the polymerase chain reaction (PCR) to detect *Mycobacterium tuberculosis* in ancient skeletons. *Int J Osteoarchaeol* **3:**137–143.
21. **Rafi A, Spiegelman M, Stanford J, Lemma E, Donoghue H, Zias J.** 1994. DNA of *Mycobacterium leprae* detected by PCR in ancient bone. *Int J Osteoarchaeol* **4:**287–290.
22. **Zink AR, Reischl U, Wolf H, Nerlich AG.** 2002. Molecular analysis of ancient microbial infections. *FEMS Microbiol Lett* **213:**141–147.

23. **Drancourt M, Raoult D.** 2008. Molecular detection of past pathogens, p 55–68. *In* Raoult D, Drancourt M (ed), *Paleomicrobiology: Past Human Infections.* Springer, Berlin–Heidelberg, Germany.

24. **Drancourt M, Raoult D.** 2005. Palaeomicrobiology: current issues and perspectives. *Nat Rev Microbiol* **3**:23–35.

25. **Dutour O.** 2011. *La paléopathologie.* Comité des Travaux Historiques et Scientifiques, Paris, France.

26. **Ruffer MA.** 1913. On pathological lesions found in Coptic bodies. *J Path Bact* **18**:149–162.

27. **Knapp M.** 2011. The next generation of genetic investigations into the Black Death. *Proc Natl Acad Sci U S A* **108**:15669–15670.

28. **Dutour O, Palfi G, Roberts C.** 2012. International congresses on the evolution and paleoepidemiology of infectious diseases, p 678–683. *In* Buikstra JE, Roberts C (ed), *The Global History of Paleopathology: Pioneers and Prospects.* Oxford University Press, New York, NY.

29. **Dutour O, Pálfi G, Bérato J, Brun J-P, ed.** 1994. *L'origine de la syphilis en Europe: avant ou après 1493?* Centre Archéologique du Var, Toulon / Errance, Paris, France.

30. **Pálfi G, Dutour O, Deák J, Hutás I, ed.** 1999. *Tuberculosis: Past and Present.* Golden Book/ Tuberculosis Foundation, Budapest/Szeged, Hungary.

31. **Roberts CA, Lewis ME, Manchester K (ed).** 2002. *The Past and Present of Leprosy. Archaeological, Historical, Paleopathological and Clinical Approaches. Proceedings of the International Congress on the Evolution and Paleoepidemiology of the Infectious Diseases 3 (ICEPID), University of Bradford, 26th-31st July, 1999.* Archaeopress, Oxford, UK.

32. **Signoli M, Chevé D, Adalian P, Boetsch G, Dutour O (ed).** 2007. *Peste: entre épidémies et sociétés / Plague: from Epidemics to Societies.* Firenze University Press, Florence, Italy.

33. **Pálfi G, Dutour O, Perrin P, Sola C, Zink A (ed).** 2015. *Tuberculosis in Evolution.* Elsevier, Edinburgh, UK.

34. **Dutour O.** 2011. Paleopathology: an archaeological approach of diseases. *TÜBA-AR* **14**:165–172.

35. **Dutour O, Signoli M, Georgeon E, Da Silva J.** 1994. Le charnier de la Grande Peste de Marseille (rue Leca): données de la fouille de la partie centrale et premiers résultats anthropologiques. *Préhistoire Anthropologie Méditerranéennes* **3**:191–204.

36. **Signoli M, Dutour O.** 1997. Le charnier des jardins du couvent de l'Observance (1722). *Provence Hist* **189**:469–488.

37. **Drancourt M, Aboudharam G, Signoli M, Dutour O, Raoult D.** 1998. Detection of 400-year-old *Yersinia pestis* DNA in human dental pulp: an approach to the diagnosis of ancient septicemia. *Proc Natl Acad Sci U S A* **95**:12637–12640.

38. **Gilbert MT, Cuccui J, White W, Lynnerup N, Titball RW, Cooper A, Prentice MB.** 2004. Absence of *Yersinia pestis*-specific DNA in human teeth from five European excavations of putative plague victims. *Microbiology* **150:** 341–354.

39. **Cunha BA.** 2004. The cause of the plague of Athens: plague, typhoid, typhus, smallpox, or measles? *Infect Dis Clin North Am* **18**:29–43.

40. **Papagrigorakis MJ, Yapijakis C, Synodinos PN, Baziotopoulou-Valavani E.** 2006. DNA examination of ancient dental pulp incriminates typhoid fever as a probable cause of the Plague of Athens. *Int J Infect Dis* **10**:206–214.

41. **Schuenemann VJ, Bos K, DeWitte S, Schmedes S, Jamieson J, Mittnik A, Forrest S, Coombes BK, Wood JW, Earn DJD, White W, Krause J, Poinar HN.** 2011. Targeted enrichment of ancient pathogens yielding the pPCP1 plasmid of *Yersinia pestis* from victims of the Black Death. *Proc Natl Acad Sci U S A* **108**:e746–e752. doi:10.1073/pnas.1105107108.

42. **Gostin LO, Berkman BE.** 2007. Pandemic influenza: ethics, law, and the public's health. *Admin Law Rev* **59**:121–175.

43. **Dutour O, Signoli M, Pálfi G.** 1998. How can we reconstruct the epidemiology of infectious diseases in the past? p 241–263. *In* Greenblatt C (ed), *Digging for Pathogens: Ancient Emerging Diseases – Their Evolutionary, Anthropological and Archaeological Context.* Balaban, Rehovot, Israel.

44. **Dutour O, Ardagna Y, Maczel M, Signoli M.** 2003. Epidemiology of infectious diseases in the past. Yersin, Koch and the skeletons, p 151–166. *In* Greenblatt C, Spiegelman M (ed), *Emerging Pathogens, Archaeology, Ecology & Evolution of Infectious Disease.* Oxford University Press, Oxford, UK.

45. **Dutour O, Maczel M, Ardagna Y.** 2007. Intérêt du "modèle peste" dans les études paléoépidémiologiques, p 89–96. *In* Signoli M, Chevé D, Adalian P, Boestch G, Dutour O (ed), *La peste: entre épidémies et sociétés.* Firenze University Press, Florence, Italy.

46. **Dutour O.** 2008. Archaeology of human pathogens: palaeopathological appraisal of palaeoepidemiology, p 125–144. *In* Raoult D, Drancourt M (ed), *Paleomicrobiology: Past Human Infections.* Springer, Berlin–Heidelberg, Germany.

47. **Bos KI, Schuenemann VJ, Golding GB, Burbano HA, Waglechner N, Coombes BK, McPhee JB, DeWitte SN, Meyer M, Schmedes S, Wood J, Earn DJD, Herring DA, Bauer P, Poinar HN, Krause J.** 2011. A draft genome of *Yersinia pestis* from victims of the Black Death. *Nature* **478**:506–510.

48. **Wagner DM, Klunk J, Harbeck M, Devault A, Waglechner N, Sahl JW, Enk J, Birdsell DN, Kuch M, Lumibao C, Poinar D, Pearson T, Fourment M, Golding B, Riehm JM, Earn DJD, DeWitte S, Rouillard J-M, Grupe G, Wiechmann I, Bliska JB, Keim PS, Scholz HC, Holmes EC, Poinar H.** 2014. *Yersinia pestis* and the Plague of Justinian 541–543 AD: a genomic analysis. *Lancet Infect Dis* **14**:319–326.

49. **Aufderheide AC, Rodriguez-Martin C.** 1998. *The Cambridge Encyclopedia of Human Paleopathology.* Cambridge University Press, Cambridge, UK.

50. **Brosch R, Gordon SV, Marmiesse M, Brodin P, Buchrieser C, Eiglmeier K, Garnier T, Gutierrez C, Hewinson G, Kremer K, Parsons LM, Pym AS, Samper S, van Soolingen D, Cole ST.** 2002. A new evolutionary scenario for the *Mycobacterium tuberculosis* complex. *Proc Natl Acad Sci U S A* **99**:3684–3689.

51. **Gutierrez MC, Brisse S, Brosch R, Fabre M, Omaïs B, Marmiesse M, Supply P, Vincent V.** 2005. Ancient origin and gene mosaicism of the progenitor of *Mycobacterium tuberculosis*. *PLoS Pathog* **1**:e5.

52. **Wirth T, Hildebrand F, Allix-Béguec C, Wölbeling F, Kubica T, Kremer K, van Soolingen D, Rüsch-Gerdes S, Locht C, Brisse S, Meyer A, Supply P, Niemann S.** 2008. Origin, spread and demography of the *Mycobacterium tuberculosis* complex. *PLoS Pathog* **4**:e1000160. doi:10.1371/journal.ppat.1000160.

53. **Kappelman J, Alcicek MC, Kazanci N, Schultz M, Ozkul M, Sen S.** 2008. First *Homo erectus* from Turkey and implications for migrations into temperate Eurasia. *Am J Phys Anthropol* **135**:110–116.

54. **Roberts CA, Pfister LA, Mays S.** 2009. Letter to the editor: was tuberculosis present in *Homo erectus* in Turkey? *Am J Phys Anthropol* **139**:442–444.

55. **Rothschild BM, Martin LD, Lev G, Bercovier H, Bar-Gal GK, Greenblatt C, Donoghue H, Spigelman M, Brittain D.** 2001. *Mycobacterium tuberculosis* complex DNA from an extinct bison dated 17,000 years before the present. *Clin Infect Dis* **33**:305–311.

56. **Hershkovitz I, Donoghue HD, Minnikin DE, Besra GS, Lee OYC, Gernaey AM, Galili E, Eshed V, Greenblatt CL, Lemma E, Bar-Gal GK, Spigelman M.** 2008. Detection and molecular characterization of 9000-year-old *Mycobacterium tuberculosis* from a Neolithic settlement in the eastern Mediterranean. *PLoS One* **3**:e3426. doi:10.1371/journal.pone. 0003426.

57. **Baker O, Lee OY, Wu HH, Besra GS, Minnikin DE, Llewellyn G, Williams CM, Maixner F, O'Sullivan N, Zink A, Chamel B, Khawam R, Coqueugniot E, Helmer D, Le Mort F, Perrin P, Gourichon L, Dutailly B, Palfi G, Coqueugniot H, Dutour O.** 2015. Human tuberculosis predates domestication in ancient Syria. *Tuberculosis (Edinb)* **95**(Suppl 1):S4–S12. doi:10.1016/j.tube.2015.02.001.

58. **Baker BJ.** 1999. Early manifestations of tuberculosis in the skeleton, p 301–307. *In* Pálfi G, Dutour O, Deák J, Hutás I (ed), *Tuberculosis: Past and Present.* Golden Book/Tuberculosis Foundation, Budapest/Szeged, Hungary.

59. **Ortner DJ, ed.** 2003. *Identification of Pathological Conditions in Human Skeletal Remains*, 2nd ed. Academic Press, San Diego, CA.

60. **Maczel M.** 2004. On the traces of tuberculosis. Diagnostic criteria of tuberculous affection in the human skeleton and their application in Hungarian and French anthropological series. PhD thesis in biological anthropology. University of La Méditerranée, Marseille, France – University of Szeged, Szeged, Hungary.

61. **Pálfi G, Bereczki Z, Ortner DJ, Dutour O.** 2012. Juvenile cases of skeletal tuberculosis from the Terry Anatomical Collection (Smithsonian Institution, Washington, D.C., USA). *Acta Biol Szeged* **56**:1–12.

62. **Haas CJ, Zink A, Molnar E, Szeimies U, Reischl U, Marcsik A, Ardagna Y, Dutour O, Palfi G, Nerlich AG.** 2000. Molecular evidence for different stages of tuberculosis in ancient bone samples from Hungary. *Am J Phys Anthropol* **113**:293–304.

63. **Ménard V, Lannelongue O.** 1888. Tuberculose vertébrale. Asselin et Houzeau, Paris, France.

64. **Sorrel É, Sorrel-Dejerine Y.** 1932. Tuberculose osseuse et ostéo-articulaire. Masson et Cie, Paris, France.

65. **World Health Organization.** 2014. Antimicrobial resistance: global report on surveillance. WHO Press, Geneva, Switzerland.

66. **Møller-Christensen V.** 1961. *Bone Changes in Leprosy.* Munksgaard, Copenhagen, Denmark.

67. **Steinbock RT.** 1976. *Paleopathological Diagnosis and Interpretation: Bone Diseases in Ancient Human Populations.* Charles C. Thomas, Springfield, IL.

68. **Robbins G, Tripathy VM, Misra VN, Mohanty RK, Shinde VS, Gray KM, Schug MD.** 2009.

Ancient skeletal evidence for leprosy in India (2000 B.C.). *PLoS One* **4**:e5669. doi:10.1371/journal.pone.0005669.

69. **Mariotti V, Dutour O, Belcastro MG, Facchini F, Brasili P.** 2005. Probable early presence of leprosy in Europe in a Celtic skeleton of the 4th–3rd century BC (Casalecchio di Reno, Bologna, Italy). *Int J Osteoarchaeol* **15**:311–325.

70. **Molto JE.** 2002. Leprosy in Roman period skeletons from Kellis 2, Dakhleh, Egypt, p 179–192. *In* Roberts CA, Lewis ME, Manchester K (ed), *The Past and Present of Leprosy: Archaeological, Historical, Paleopathological and Clinical Approaches. Proceedings of the International Congress on the Evolution and Paleoepidemiology of the Infectious Diseases 3 (ICEPID), University of Bradford, 26th-31st July 1999.* Archaeopress, Oxford, UK.

71. **Mark S.** 2002. Alexander the Great, seafaring, and the spread of leprosy. *J Hist Med Allied Sci* **57**:285–311.

72. **Donoghue HD, Marcsik A, Matheson C, Vernon K, Nuorala E, Molto JE, Greenblatt CL, Spigelman M.** 2005. Co-infection of *Mycobacterium tuberculosis* and *Mycobacterium leprae* in human archaeological samples: a possible explanation for the historical decline of leprosy. *Proc Biol Sci* **272**:389–394.

73. **Minnikin DE, Besra GS, Lee O-YC, Spigelman M, Donoghue HD.** 2011. The interplay of DNA and lipid biomarkers in the detection of tuberculosis and leprosy in mummies and other skeletal remains, p 109–114. *In* Gill-Frerking H, Rosendahl W, Zink A, Piombini-Mascali D (ed), *Yearbook of Mummy Studies*, vol 1. Verlag Dr. Friedrich Pfeil, Münich, Germany.

74. **Monot M, Honore N, Garnier T, Zidane N, Sherafi D, Paniz-Mondolfi A, Matsuoka M, Taylor GM, Donoghue HD, Bouwman A, Mays S, Watson C, Lockwood D, Khamesipour A, Dowlati Y, Jianping S, Rea TH, Vera-Cabrera L, Stefani MM, Banu S, Macdonald M, Sapkota BR, Spencer JS, Thomas J, Harshman K, Singh P, Busso P, Gattiker A, Rougemont J, Brennan PJ, Cole ST.** 2009. Comparative genomic and phylogeographic analysis of *Mycobacterium leprae*. *Nat Genet* **41**:1282–1289.

75. **Schuenemann VJ, Singh P, Mendum TA, Krause-Kyora B, Jager G, Bos KI, Herbig A, Economou C, Benjak A, Busso P, Nebel A, Boldsen JL, Kjellstrom A, Wu H, Stewart GR, Taylor GM, Bauer P, Lee OY, Wu HH, Minnikin DE, Besra GS, Tucker K, Roffey S, Sow SO, Cole ST, Nieselt K, Krause J.** 2013. Genome-wide comparison of medieval and modern *Mycobacterium leprae*. *Science* **341**:179–183.

76. **Mendum T, Schuenemann V, Roffey S, Taylor G, Wu H, Singh P, Tucker K, Hinds J, Cole S, Kierzek A, Nieselt K, Krause J, Stewart G.** 2014. *Mycobacterium leprae* genomes from a British medieval leprosy hospital: towards understanding an ancient epidemic. *BMC Genomics* **15**:e270. doi:10.1186/1471-2164-15-270.

77. **Donoghue HD, Michael Taylor G, Marcsik A, Molnar E, Palfi G, Pap I, Teschler-Nicola M, Pinhasi R, Erdal YS, Veleminsky P, Likovsky J, Belcastro MG, Mariotti V, Riga A, Rubini M, Zaio P, Besra GS, Lee OY, Wu HH, Minnikin DE, Bull ID, O'Grady J, Spigelman M.** 2015. A migration-driven model for the historical spread of leprosy in medieval Eastern and Central Europe. *Infect Genet Evol* **31**:250–256.

78. **Wiechmann I, Grupe G.** 2005. Detection of *Yersinia pestis* DNA in two early medieval skeletal finds from Aschheim (Upper Bavaria, 6th century A.D.). *Am J Phys Anthropol* **126**:48–55.

79. **Raoult D, Aboudharam G, Crubezy E, Larrouy G, Ludes B, Drancourt M.** 2000. Molecular identification by "suicide PCR" of *Yersinia pestis* as the agent of medieval black death. *Proc Natl Acad Sci U S A* **97**:12800–12803.

80. **Yersin A.** 1894. La peste bubonique à Hong-Kong. *Ann Inst Pasteur (Paris)* **8**:662–667.

11

Past Bartonelloses

PIERRE-EDOUARD FOURNIER[1]

To date, several bacteria have been detected in ancient specimens as causative agents of various diseases, in both humans and animals (1). These include *Bartonella henselae* (2), *Bartonella quintana* (3), *Borrelia burgdorferi* (4), *Enterobacteriaceae* (5), *Mycobacterium leprae* (6), *Mycobacterium tuberculosis* (7), *Rickettsia prowazekii* (8), *Treponema pallidum*, and *Yersinia pestis* (9). Most of these microorganisms have been detected with molecular methods (1), although culture, immunohistochemistry (10), and the detection of specific antibodies (11) have occasionally been successful.

Bartonella species are Gram-negative, facultatively intracellular bacteria belonging to the *Alphaproteobacteria* phylum. These microorganisms are zoonotic agents, associated with various arthropod vectors and infecting mammals, including humans. Members of the *Bartonella* genus were first detected in the early 1900s by Alberto Barton, a Peruvian physician who studied patients afflicted with Oroya fever (12), and *Bartonella bacilliformis*, the agent of the disease, was described in 1915 (13).

To date, *Bartonella* species have been identified in a variety of ancient specimens from diverse periods. The paleomicrobiological evidence of *Bartonella* infections is summarized in Table 1.

[1]Unité de Recherche sur les Maladies Infectieuses et Tropicales Emergentes (URMITE), UM63, CNRS7278, IRD198, Inserm 1095, Aix-Marseille Université, Marseille, France.
Paleomicrobiology of Humans
Edited by Michel Drancourt and Didier Raoult
© 2016 American Society for Microbiology, Washington, DC
doi:10.1128/microbiolspec.PoH-0007-2015

TABLE 1 Ancient specimens in which *Bartonella* species have been detected

Specimen and/or body site	Source	Datation	Location	Identified species	Detection method	Study date	Reference
Coprolite	Human	14th century	Belgium	*B. quintana, B. henselae, B. tribocorum*	Metagenomics	2014	21
Dental pulp	Cat	13th–18th centuries	France	*B. henselae*	PCR (*groEL, pap31*)	2004	2
Dental pulp	Human	4,000 years old	France	*B. quintana*	PCR (*groEL, hbpE*)	2005	3
Dental pulp	Human	11th–15th centuries	France	*B. quintana, Yersinia pestis*	PCR (16S-23S rRNA, *pla*)	2011	19
Dental pulp	Human	Napoleon's Grand Army Russian retreat, 1812	Lithuania	*B. quintana*	PCR (*hbpE, htrA*)	2006	8
Skin biopsies	Human mummy	10th century	Peru	*Bartonella*-like bacteria	Microscopy	1974	27

IDENTIFICATION OF *BARTONELLA* SPECIES IN ANCIENT SPECIMENS

Bartonella quintana

B. quintana was first described in 1917 (14) as *Rickettsia quintana*, the agent of trench fever, an incapacitating and relapsing bacteremia that affected British soldiers during World War I (15). The bacterium is transmitted among humans, who constitute its reservoir, by the body louse, *Pediculus humanus humanus*; it has also been given such various names as *Rickettsia pediculi*, *Rickettsia wolhynica*, *Rickettsia weigli*, *Burnetia wolhynica*, *Wolhynia quintanae*, and *Rochalimaea quintana*. In 1993, it was reclassified as *B. quintana* following unification of the genera *Rochalimaea* and *Bartonella* (16). In addition to trench fever, *B. quintana* is an agent of chronic bacteremia and endocarditis in homeless people, and of bacillary angiomatosis in immunocompromised patients (17).

The oldest evidence of *B. quintana*, and the oldest evidence of a *Bartonella* species, was obtained in an individual buried 4,000 years ago (3). In this study, in which a "suicide PCR" (i.e., a nested PCR assay targeting for the first time in the laboratory a fragment of the *hbpE* gene) was used (18), and in which no positive control but many negative controls were incorporated, the authors obtained a positive amplification from the pulp of one molar from one individual buried in Peyraoutes, in southeastern France, and dated by radiocarbon to between 2230 and 1950 BC. Following sequencing, the amplicon was identified as *B. quintana*, which was confirmed by partial amplification and sequencing of the *groEL* gene (3). This result demonstrated the old association of *B. quintana* with humans.

In 2007, three multiple burial sites were discovered in Bondy, France. A total of 11 skeletons were found and dated from the 11th to the 15th centuries (19). By amplifying the *pla* gene and the 16S to 23S rRNA spacer from the dental pulp of five of these individuals, Tran et al. detected *Yersinia pestis* genotype Orientalis and *B. quintana* in two and three of them, respectively. This co-infection may be explained by the occurrence of a plague outbreak in a population exposed to body lice infected with *B. quintana*, or the transmission of both pathogens by lice, as previously suggested (20).

In addition, *B. quintana* was detected in a 14th century coprolite found inside a barrel at a medieval site in Namur, Belgium, in 1996 (21). Following acridine orange binding of DNA in the coprolite, the authors used V6

16S rRNA region-based metagenomics to decipher the microbiota of this specimen. Among the 107,470 sequencing reads obtained, 53 reads of 270 to 375 bp were from *B. quintana* (21), suggesting that the local population was at least partially infected by louse-associated pathogens.

Finally, a search for louse-transmitted pathogens in the remains of soldiers of Napoleon's Grand Army, found in a mass grave in Vilnius, Lithuania, in 2001, enabled the identification of *B. quintana* in three body louse fragments and in the pulp of 10 teeth from seven soldiers. Results of suicide PCR partially targeting the *hbpE* gene were confirmed by a second, *htrA*-specific assay (8). This study confirmed the supposed role of louse-borne pathogens during the French retreat from Russia.

Bartonella henselae

B. henselae was first detected in 1990, by 16S rRNA PCR and sequencing, in skin biopsy specimens of patients with bacillary angiomatosis (22). The bacterium was formally described in 1992 as *Rochalimaea henselae* (23). In 1993, it was reclassified within the *Bartonella* genus as *B. henselae* (16). This species is an agent of chronic bacteremia in felids, which are its reservoir. Domestic cats transmit *B. henselae* to humans via scratches (transmission by bites has not been fully proven) and cause a variety of diseases that range from the frequent cat-scratch disease to rarer forms, such as endocarditis, bacillary angiomatosis, and peliosis hepatis (17).

In 2004, La et al. detected *B. henselae* in the dental pulp of three cats dating from the 13th, 14th, and 16th centuries that were found in burial sites in France (medieval Louvre in Paris, Montbéliard, and Amiens, respectively) (2). These results were obtained by using two independent suicide PCR assays targeting the *pap*31 and *groEL* genes. This study was the first paleomicrobiological demonstration of bartonellae in their animal reservoir and showed that *B. henselae* had been associated

with cats for at least 800 years. In addition, *B. henselae* was detected in the previously mentioned 14th century coprolite found in Namur, within which 47 reads of 203 to 251 bp were specific for this species (21). These data suggested that coprolites are useful for identifying ancient human pathogens, including zoonotic agents.

Bartonella bacilliformis

B. bacilliformis is the agent of Carrión's disease, a potentially lethal and biphasic infection characterized by an acute bacteremia causing hemolytic anemia (Oroya fever) and a chronic, indolent, eruptive skin disease (verruga peruana). The disease is transmitted to humans by the sand fly, *Lutzomyia verrucarum*, in the Andes Mountains from Chile to Colombia (24).

Verruga peruana was first depicted on Ecuadorian huacas (ceramic figurines) dated to 1,000 years before the discovery of the New World by Columbus (25) and was later described by Pizzaro in 1571 (26). Oroya fever was also probably recognized by pre-Columbian Indians, as suggested by the word *sirki* ("anemia") in the Quechua language (24). However, it was not before the early 20th century that its causative agent was observed by Alberto Barton in the red blood cells of patients afflicted with Carrión's disease (12). To date, the only evidence of *B. bacilliformis* in ancient specimens was reported in 1974 (27). The authors observed bacterial clusters that were morphologically compatible with bartonellae in skin biopsy specimens from the verruga peruana lesions of a male mummy found in a burial gallery in southern Peru in 1960. This mummy, from Huari culture, dated from the 10th century (27).

Other Bartonella Species

In the metagenomic analysis of the 14th century coprolite found in Belgium, Appelt et al. also obtained one 306-bp read whose best match was *B. tribocorum*, a rat-associated species of unknown pathogenicity for humans (21).

CONCLUSION

Paleomicrobiological studies have demonstrated that *Bartonella* species are human pathogens that are not only of current concern; they have long been associated with humans and/or their pets in various areas of the world. There is strong evidence that these bacteria have played a role in famous historical episodes, such as the conquest of South America by the Spanish and the retreat from Russia by the French soldiers of Napoleon's Army.

CITATION

Fournier P-E. 2016. Past bartonelloses. Microbiol Spectrum 4(3):PoH-0007-2015.

REFERENCES

1. **Drancourt M, Raoult D.** 2005. Palaeomicrobiology: current issues and perspectives. *Nat Rev Microbiol* **3:**23–35.
2. **La VD, Clavel B, Lepetz S, Aboudharam G, Raoult D, Drancourt M.** 2004. Molecular detection of *Bartonella henselae* DNA in the dental pulp of 800-year-old French cats. *Clin Infect Dis* **39:**1391–1394.
3. **Drancourt M, Tran-Hung L, Courtin J, Lumley H, Raoult D.** 2005. *Bartonella quintana* in a 4000-year-old human tooth. *J Infect Dis* **191:**607–611.
4. **Marshall WF III, Telford SR III, Rys PN, Rutledge BJ, Mathiesen D, Malawista SE, Spielman A, Persing DH.** 1994. Detection of *Borrelia burgdorferi* DNA in museum specimens of *Peromyscus leucopus*. *J Infect Dis* **170:**1027–1032.
5. **Rhodes AN, Urbance JW, Youga H, Corlew-Newman H, Reddy CA, Klug MJ, Tiedje JM, Fisher DC.** 1998. Identification of bacterial isolates obtained from intestinal contents associated with 12,000-year-old mastodon remains. *Appl Environ Microbiol* **64:**651–658.
6. **Spigelman M, Donoghue HD.** 2001. Brief communication: unusual pathological condition in the lower extremities of a skeleton from ancient Israel. *Am J Phys Anthropol* **114:**92–93.
7. **Rothschild BM, Martin LD, Lev G, Bercovier H, Bar-Gal GK, Greenblatt C, Donoghue H, Spigelman M, Brittain D.** 2001. *Mycobacterium tuberculosis* complex DNA from an extinct bison dated 17,000 years before the present. *Clin Infect Dis* **33:**305–311.
8. **Raoult D, Dutour O, Houhamdi L, Jankauskas R, Fournier PE, Ardagna Y, Drancourt M, Signoli M, La VD, Macia Y, Aboudharam G.** 2006. Evidence for louse-transmitted diseases in soldiers of Napoleon's Grand Army in Vilnius. *J Infect Dis* **193:**112–120.
9. **Drancourt M, Roux V, Dang LV, Lam THCD, Chenal-Francisque V, Ogata H, Fournier PE, Crubezy E, Raoult D.** 2004. Genotyping, Orientalis-like *Yersinia pestis*, and plague pandemics. *Emerg Infect Dis* **10:**1585–1592.
10. **Dumler JS, Baisden BL, Yardley JH, Raoult D.** 2003. Immunodetection of *Tropheryma whipplei* in intestinal tissues from Dr. Whipple's 1907 patient. *N Engl J Med* **348:**1411–1412.
11. **Kolman CJ, Centurion-Lara A, Lukehart SA, Owsley DW, Tuross N.** 1999. Identification of *Treponema pallidum* subspecies *pallidum* in a 200-year-old skeletal specimen. *J Infect Dis* **180:**2060–2063.
12. **Barton A.** 1901. El Germen Patógeno de la Enfermedad de Carrión. *La Crónica Médica* **XVIII:**209–216.
13. **Strong RP, Tyzzer EE, Sellards AW.** 1915. Oroya fever. Second report. *J Am Med Assoc* **64:**806–808.
14. **Schminke A.** 1917. Histopatologischer Befund in Roseolen der Haut bei wolhynischem Fieber. *Münch Med Wochenschr* **64:**961.
15. **Byam W, Caroll JH, Churchill JH, Dimond L, Sorapure VE, Wilson RM, Lloyd LL.** 1919. *Trench fever – a louse-borne disease.* Oxford University Press, London, UK.
16. **Brenner DJ, O'Connor S, Winkler HH, Steigerwalt AG.** 1993. Proposals to unify the genera *Bartonella* and *Rochalimaea*, with descriptions of *Bartonella quintana* comb. nov., *Bartonella vinsonii* comb. nov., *Bartonella henselae* comb. nov., and *Bartonella elizabethae* comb.nov., and to remove the family *Bartonellaceae* from the order *Rickettsiales*. *Int J Syst Bacteriol* **43:**777–786.
17. **Angelakis E, Raoult D.** 2014. Pathogenicity and treatment of *Bartonella* infections. *Int J Antimicrob Agents* **44:**16–25.
18. **Raoult D, Aboudharam G, Crubezy E, Larrouy G, Ludes B, Drancourt M.** 2000. Molecular identification by "suicide PCR" of *Yersinia pestis* as the agent of medieval black

death. *Proc Natl Acad Sci U S A* **97**:12800–12803.

19. **Tran TN, Forestier CL, Drancourt M, Raoult D, Aboudharam G.** 2011. Brief communication: co-detection of *Bartonella quintana* and *Yersinia pestis* in an 11th-15th burial site in Bondy, France. *Am J Phys Anthropol* **145**:489–494.

20. **Ayyadurai S, Sebbane F, Raoult D, Drancourt M.** 2010. Body lice, *Yersinia pestis* Orientalis, and black death. *Emerg Infect Dis* **16**:892–893.

21. **Appelt S, Armougom F, Le Bonhomme M, Robert C, Drancourt M.** 2014. Polyphasic analysis of a Middle Ages coprolite microbiota, Belgium. *Plos One* **9**:e88376. doi:10.1371/journal.pone.0088376.

22. **Relman DA, Loutit JS, Schmidt TM, Falkow S, Tompkins LS.** 1990. The agent of bacillary angiomatosis: an approach to the identification of uncultured pathogens. *N Engl J Med* **323**:1573–1580.

23. **Regnery RL, Anderson BE, Clarridge JE, Rodriguez-Barradas MC, Jones DC, Carr JH.** 1992. Characterization of a novel *Rochalimaea* species, *R. henselae* sp. nov., isolated from blood of a febrile, human immunodeficiency virus-positive patient. *J Clin Microbiol* **30**:265–274.

24. **Minnick MF, Anderson BE, Lima A, Battisti JM, Lawyer PG, Birtles RJ.** 2014. Oroya fever and verruga peruana: bartonelloses unique to South America. *PLoS Negl Trop Dis* **8**:e2919. doi:10.1371/journal.pntd.0002919.

25. **Alexander B.** 1995. A review of bartonellosis in Ecuador and Colombia. *Am J Trop Med Hyg* **52**:354–359.

26. **Prescott WH.** 1998. *History of the Conquest of Peru.* Modern Library, New York, NY.

27. **Allison MJ, Pezzia A, Gerszten E, Mendoza D.** 1974. A case of Carrion's disease associated with human sacrifice from the Huari culture of Southern Peru. *Am J Phys Anthropol* **41**:295–300.

Paleomicrobiology of Human Tuberculosis

12

HELEN D. DONOGHUE[1]

THE MODERN DISEASE

Tuberculosis remains one of the world's deadliest communicable diseases. In 2014, tuberculosis developed in an estimated 9.6 million people, and 1.5 million died of the disease (1). The principal causative organism is *Mycobacterium tuberculosis*, an obligate pathogen that is a member of the *M. tuberculosis* complex (MTBC), a group of closely related organisms that primarily infect different animal hosts. Tuberculosis may involve every organ in the body, but the most common clinical presentation is pulmonary disease, in which transmission is via infectious aerosols released from the lungs of an infected person. In the alveolus of the lung, inhaled tubercle bacilli are ingested by macrophages and are normally contained by the host immune response. This leads to granuloma formation and eventually to calcified lesions. Swallowing infected sputum can cause intestinal tuberculosis. Transmission can occur via direct contact in cases of scrofula (skin tuberculosis). In addition, ingestion of milk or food from an infected animal can cause human infection with *Mycobacterium bovis* or other members of the MTBC. However, subsequent transmission of these animal MTBC lineages from person to person is rare. *M. tuberculosis*

[1]Centre for Clinical Microbiology, Division of Infection and Immunity, University College London, United Kingdom.
Paleomicrobiology of Humans
Edited by Michel Drancourt and Didier Raoult
© 2016 American Society for Microbiology, Washington, DC
doi:10.1128/microbiolspec.PoH-0003-2014

can survive and grow within macrophages, so that it is able to evade the host immune system. An active cell-mediated immune response is required to contain and kill the tubercle bacilli, so any underlying conditions that reduce its efficiency increase susceptibility to tuberculosis. One-third of the global population is estimated to have latent tuberculosis infection. These individuals do not have active disease but may develop it in the near or remote future, a process called tuberculosis reactivation. The lifetime risk for reactivation is estimated to be 5% to 10%, with tuberculosis developing in the majority of cases within the first 5 years after initial infection. However, the risk is considerably higher in the presence of predisposing factors (2).

PALEOPATHOLOGY OF TUBERCULOSIS

Skeletal Changes Indicative of Tuberculosis

The most characteristic visible skeletal changes in archaeological cases of tuberculosis are those to the spine, such as Pott's disease (Fig. 1A) and cold (chronic) abscess (Fig. 1B). Pott's disease is diagnosed by characteristic changes that result in kyphosis, or gibbus, in which there is loss of function in the lower limbs due to damage to the spinal column. Tuberculosis can affect any part of the skeleton, but bony joints are common sites of involvement. Changes associated with tuberculosis are periosteal reactive lesions on tubular bones, hypertrophic ostearthropathy, and osteomyelitis (3, 4). It is estimated that approximately 40% of cases of skeletal tuberculosis result in tuberculosis of the spine (5). However, as skeletal tuberculosis occurs in only 3% to 5% of untreated cases, the incidence of tuberculosis in the past was undoubtedly far higher than that suggested by the number of bony lesions observed (6). Historical texts contain recognizable descriptions of tuberculosis, in which it is identified as phthisis, scrofula, King's Evil, lupus vulgaris, or consumption, for example (7). Detailed morphological studies enabled diagnostic criteria to be agreed upon, based on more recent historical skeletal collections with contemporaneous records of individual cases, including age, sex, occupation, symptoms, and cause of death (8–10). It was noted that periostitis (surface changes caused by new bone formation) on ribs was significantly associated with individuals in whom clinical tuberculosis had been diagnosed (Fig. 1C). Other conditions linked to recognized tuberculosis changes include hypertrophic ostearthropathy (11, 12) and serpens endocrania symmetrica—a morphological sign of respiratory distress and increased vascularization around the brain (11).

Archaeological Reports of Tuberculosis around the World

Paleopathology suggestive of tuberculosis has been reported from predynastic Egypt (3500 to 2650 BC) (13, 14), middle Neolithic Italy at the beginning of the fourth millennium BC (15), and an eastern Mediterranean Pre-Pottery Neolithic site (9250 to 8160 years BP) (16). There are fewer reports from eastern and southeastern Asia, but tuberculosis was present in northeastern Thailand at an Iron Age site dated from 2500 to 1700 years BP (17) and in Japan and Korea at least 2,000 years ago (18). Precolonial tuberculosis in the Americas was first identified in humans in a mummified child with bone pathology suggestive of tuberculosis, dated to approximately 700 AD, from the Nazca culture of southern Peru (19, 20). It was also recognized in northwestern Argentina (21) and northern Chile (22), with most morphological evidence found in the period from 500 to 1000 AD, corresponding to fully agropastoral societies. More recently, tuberculosis has been confirmed in Peru from Chiribaya cultures (750 to 1350 AD) associated with the Middle Horizon/Late Intermediate Period (23).

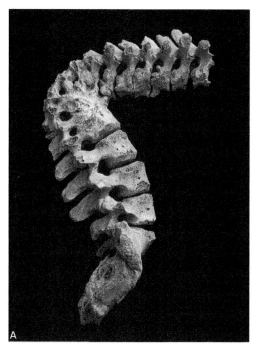

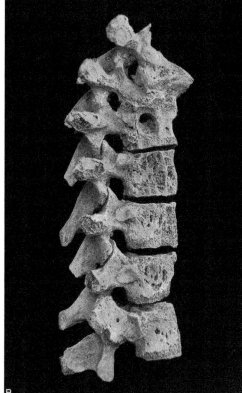

FIGURE 1 (A) Paleopathology diagnostic for skeletal tuberculosis: Pott's disease, angular kyphosis in Th8–L2. Hungary: Zalavár-Vársziget-Kápolna, juvenile, grave No. 17/03. (B) Paleopathology highly suggestive of tuberculosis: evidence of infection shown by fusion of vertebrae (Th6–8) with slight gibbus, cavities, and traces of cold abscess (chronic lytic lesion). Hungary: Zalavár-Vársziget-Kápolna, juvenile, grave No. 74/03. (C) Paleopathology showing nonspecific changes consistent with a tuberculosis infection; disseminated, small, new bone formations can be observed on the costal groove and on the inner surface of the ribs. Romania: Peteni, grave No. 107. (Courtesy of Tamás Hadju, Department of Biological Anthropology, Eötvös Loránd University, Budapest, Hungary. Fig. 1A, B reprinted from *HOMO - Journal of Comparative Human Biology* [95] with permission of the publisher. Fig. 1C reprinted from *Spine* [96] with permission of the publisher.)

Relationship of Tuberculosis to Early Human Populations

Because *M. tuberculosis* is an obligate pathogen with no environmental reservoir, its per-sistence is related to the density of the human population. Therefore, the long hunter–gatherer stage of human evolution, consisting of small populations, would select for commensal organisms or for pathogens that could be transmitted decades after infecting a host, after new susceptible individuals had been introduced into the population via births or migration (24). Typically, commensals are transmitted vertically from parent to child, whereas pathogens are transmitted horizontally. However, tuberculosis is an intermediate case because in a low-density population individuals are more likely to spread infection to family members than to strangers.

The Neolithic transition and development of agriculture were associated with a pronounced increase in tuberculosis prevalence (25, 26). Indirect evidence of this association between urbanization and tuberculosis is the relationship between human natural resistance to the disease and long-term urban settlements (27). Although a majority of individuals have a long or lifetime tuberculosis infection, disease may be latent or have phases of activity, which then subside. Pathogen and host can co-exist, which provides a reservoir of infection for the pathogen and may cause selection pressure on the survival of its human host. Early in life, there is the opportunity for tuberculosis transmission, as infants with an immature cell-mediated immune system can develop active disease with a high mortality rate. Late transmission can occur when adults become susceptible from causes that increase their susceptibility, such as malnutrition, warfare, and old age. The transition from foraging to settled farming communities in the Neolithic period coincided with the appearance of diseases associated with larger, denser populations, a sedentary lifestyle, widespread domestication of animals, social stratification, and a less varied diet (28, 29).

Agriculture in the Old World is evident from about 10,000 years ago, where five independent areas of cultivation emerged in Mesopotamia, sub-Saharan Africa, southeastern Asia, northern China, and southern China. Initially, it was believed that humans acquired tuberculosis from animals, especially after domestication (30), because this coincided with the observed human paleo-pathology. As we now know that the human tubercle bacillus is of a more ancestral lineage (31), it is likely that animal domestication was important in sustaining a denser human population, thereby enabling tuberculosis to become endemic (16). However, it is most unlikely that the bovine tuberculosis lineage was derived from the lineage that principally infects humans (32).

DETECTION AND MOLECULAR DIAGNOSIS OF TUBERCULOSIS

The traditional method of diagnosis, still used in many parts of the world today, is chest radiology plus the microscopic examination of sputum smears following Ziehl-Neelsen staining. This method identifies only 10% to 30% of cases, even when enhanced by fluorescence microscopy, so diagnosis is confirmed by culture in solid or liquid media. Because *M. tuberculosis* can take 4 to 6 weeks to grow, the World Health Organization recommends rapid diagnostic methods based on *M. tuberculosis* genetic markers and PCR, such as the Xpert MTB/RIF rapid TB test, for the diagnosis of pulmonary and extrapulmonary tuberculosis in adults and children (33). It was the early development of *M. tuberculosis* molecular diagnostic markers that led to the discovery of tuberculosis in archaeological material.

Detection of Archaeological and Historical *Mycobacterium tuberculosis* Ancient DNA

M. tuberculosis was first identified in the pre-Columbian Americas by using tissue from a mummified child from the Nazca culture of southern Peru (19, 20). As previously described, this mummy had bone pathology suggestive of tuberculosis and microscopic evidence of acid-alcohol–resistant bacilli. Although these findings are highly suggestive of active tuberculosis, molecular evidence was required to confirm the diagnosis.

Characteristics of Ancient DNA

Modern DNA sequences will outnumber ancient DNA (aDNA) in any sample, so stringent precautions must be taken, throughout the excavation and sampling process, to reduce extraneous contamination to a minimum. In living cells, DNA is subjected to enzymatic repair processes, but after death DNA is rapidly degraded by enzymes derived from both the host and the macro and microbial flora that form part of the natural decay process (34). As a result of cumulative changes over time (diagenesis), aDNA may develop hydrolytic and oxidative lesions. The breakdown of the N-glycosyl bond between the sugar and the base, in the presence of water, leads to hydrolytic cleavage and DNA fragmentation. Hydrolytic depurination causes a preferential loss of guanine and adenine, whereas the pyrimidines cytosine and thymine are 40-fold more susceptible to hydrolytic deaminization (35). Oxidative damage, especially to pyrimidines, can result in the formation of substances such as hydantoins, which block extension during PCR (36). DNA strands may also become chemically cross-linked as a result of the formation of Maillard products (37) by condensation reactions between sugars and primary amino groups in proteins and nucleic acids (34). Local environmental conditions have a strong impact on the persistence of aDNA, such as the temperature, the pH at the site, the availability of water and oxygen, and the fluctuations of all these factors over time (38). Indeed, these factors outweigh the impact of the chronological age of samples.

Mycobacterial DNA is more robust than the DNA of mammals (39), but its persistence depends not only upon the local environmental conditions but also on the nature of the infection at the time of death of its host. Therefore, *M. tuberculosis* aDNA is often highly localized, and DNA extraction protocols may have to be optimized for specimens from different sites (40). DNA extraction normally involves the disaggregation of samples with ethylenediamine-tetraacetic acid (EDTA) and proteinase K. Covalent cross-links can be reduced by the reagent N-phenacylthiazolium bromide (PTB), which cleaves glucose-derived protein cross-links (37). The final stage is disruption of samples with lysis buffer based on guanidium thiocyanate or hydrochloride, followed by silica capture or isopropanol precipitation of aDNA, washing, and drying.

Methods of *Mycobacterium tuberculosis* Complex Ancient DNA Analysis

The MTBC was one of the first groups of microorganisms to benefit from the introduction of molecular diagnostics because of their very slow growth rate and clinical significance. Early molecular detection of *M. tuberculosis* was based on short palindromic repeat sequences. Insertion sequences IS*6110* and IS*1081* were identified as useful specific targets for PCR analysis (41). IS*6110* ranges from 1 to 24 copies per cell but is absent in rare strains from southeastern Asia (42), whereas IS*1081* is present at 1 copy per cell (41, 43) and so can be used for quantitative analysis.

Initially, conventional PCR was used to detect ancient and historical tuberculosis, followed by agarose gel electrophoresis for the detection of amplicons. Because of the tendency of aDNA to fragment, there should be an inverse correlation between the length of the target sequence and amplification efficiency, with claims of long amplicons subject to scrutiny. Results should be repeated in a second extract and verified in an independent laboratory. The use of real-time PCR, based on specific primers and fluorescent probes enables shorter DNA fragments to be examined.

Verification of *Mycobacterium tuberculosis* Complex Ancient DNA Findings

Initially, there was considerable skepticism among anthropologists when tuberculosis was reported in skeletal material with non-specific or no paleopathology, although to clinical microbiologists, the findings were unsurprising. Suggested criteria for analysis were based on host aDNA, in which protein preservation was used as a marker to indicate the likelihood of successful detection of aDNA, although this relationship has since been questioned (44). In any event, because of the thick, lipid-rich bacterial cell wall and the DNA high guanine–cytosine (GC) content, mycobacterial aDNA is more persistent than the surrounding host aDNA (39), so such prior screening is unnecessary.

In the early days of aDNA research, there were genuine concerns about the prevention of cross-contamination between samples and amplified DNA. For work on host DNA, stringent containment facilities with one-way access and negative air pressure have been designed to minimize the possibility of contamination with modern DNA or amplicons. Although careful precautions are required, work on the MTBC can be accomplished with the use of good microbiological technique and the strict separation of different stages of DNA extraction, amplification, and subsequent analysis (45, 46). This is because the organisms are pathogens with no known environmental reservoir.

Mycobacterium tuberculosis Complex Genotypes, Strains, and Lineages

PCR-based typing methods have facilitated epidemiological studies of tuberculosis. An early example is spoligotyping, which is based on the direct repeat (DR) region of the MTBC (47). PCR primers are used to amplify 43 unique spacer regions that lie between each DR locus, and amplicons from individual spacers are visualized by dot–blot hybridization on a membrane. Spoligotyping and typing based on other repetitive elements clearly distinguish members of the MTBC and can identify different lineages. *M. tuberculosis* strains commonly show deletions, and because the loss of spacers is unidirectional, the

data can indicate evolutionary trends (31, 48). Synonymous single-nucleotide polymorphisms (SNPs) or variants (SNVs) are functionally neutral and so can also be used to distinguish between lineages, aided by the virtual lack of horizontal gene transfer. This has led to the recognition of seven phylogeographical lineages (Fig. 2), each associated with specific human populations (49–52), with the animal lineages sometimes described as lineage 8 (53). High-throughput sequencing of entire genomes, coupled with updates in bioinformatics analysis, is the latest tool used to elucidate the relationships between lineages and strains. Recent genomic analyses suggest that *M. tuberculosis* has evolved from a pool of smooth colony-like mycobacteria (STMs) that gained additional virulence and persistence mechanisms, including loss of gene function, acquisition of new genes via horizontal gene transfer, interstrain recombination of gene clusters, and fixation of SNPs (54). The individual members of the MTBC (excluding the STMs classified as *Mycobacterium canettii*) are 99.95% identical on the basis of nucleotide sequence. This has led to the suggestion that there was an evolutionary bottleneck at the time of speciation. The estimated date of this event (Fig. 3) varies from 3 million years ago if the STMs are included (48) to 40,000 years ago (55), to 70,000 years ago (56), to only 6,000 years ago—based on pre-European contact Peruvian material (23). Clearly, the identification of the Most Recent Common Ancestor (MRCA) is crucial in such calculations (32).

Mycobacterium tuberculosis Cell Wall Lipid Biomarkers

In parallel with *M. tuberculosis* aDNA studies, the use of specific mycobacterial cell wall lipid biomarkers has been developed. *M. tuberculosis* has a cell envelope incorporating a peptidoglycan-linked arabinogalactan esterified by long-chain mycolic acids. A range of "free" lipids is associated with the "bound" mycolic acids, producing

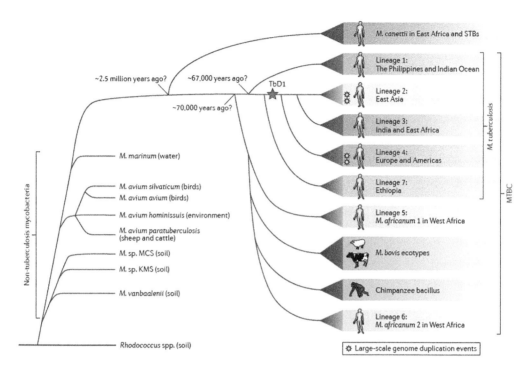

FIGURE 2 Evolutionary relationship between selected mycobacteria and members of the *Mycobacterium tuberculosis* complex (MTBC). The MTBC was thought to arise as a clonal expansion from a smooth tubercle bacillus (STB) progenitor population. The animal-adapted *Mycobacterium bovis* ecotypes branch from a presumed human-adapted lineage of *Mycobacterium africanum* that is currently restricted to West Africa. Human-adapted *M. tuberculosis* strains are grouped into seven main lineages, each of which is primarily associated with a distinct geographical distribution. The dates of branching events are only crude estimates. (Courtesy of James E. Galaghan, Department of Biomedical Engineering, Bioinformatics Program and National Emerging Infectious Diseases Laboratory, Boston University, Boston, Massachusetts, USA, and Broad Institute of Massachusetts Institute of Technology and Harvard, Cambridge, Massachusetts, USA. Reprinted from *Nature Reviews Genetics* [97] with permission of the publisher.)

an effective envelope outer membrane. The distribution of these lipids varies among mycobacteria, and such lipids can act as specific biomarkers in the identification of *M. tuberculosis* and in tracing its evolution (40, 57). The advantage of lipid biomarkers is that they are detected by extremely sensitive methods, so that there is no amplification of material. Initially, detection of the 70 to 90 carbon mycolic acids was used to complement DNA amplification and paleopathology (58, 59). The biomarker range now includes multi-methyl-branched mycocerosic and mycolipenic acids (Fig. 4A, B) (40, 60). Mycolic acids were originally analyzed by

fluorescence high performance liquid chromatography (HPLC) of slightly unstable methylanthryl esters (59), so a special robust derivatization protocol, involving pyrenebutyrates of pentafluorobenzyl (PFB) esters, was systematically developed (16, 59, 60). Selected ion monitoring (SIM) negative ion-chemical ionization gas chromatography mass-spectrometry (NICI-GCMS) is an exquisitely sensitive detection method for the mycocerosate and mycolipenate PFB esters (60–63).

With the aim of limiting destructive analyses, it is useful to know that the aqueous residues from DNA extractions can be used

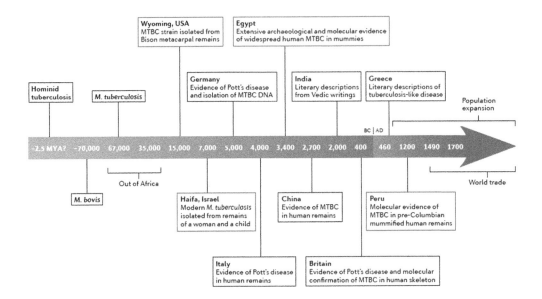

FIGURE 3 A possible timeline of evolutionary events and archaeological data; the location for archaeological evidence is indicated in each box. Boxes outlined in black indicate morphological evidence only, whereas boxes outlined in red denote both morphological and molecular evidence. (Courtesy of James E. Galaghan, Department of Biomedical Engineering, Bioinformatics Program and National Emerging Infectious Diseases Laboratory, Boston University, Boston, Massachusetts, USA, and Broad Institute of Massachusetts Institute of Technology and Harvard, Cambridge, Massachusetts, USA. Reprinted from *Nature Reviews Genetics* [97] with permission of the publisher.)

for lipid extractions because these use hydrophobic reagents that release different components from samples (64).

MYCOBACTERIUM TUBERCULOSIS FINDINGS BASED ON MOLECULAR BIOMARKERS

Overview of *Mycobacterium tuberculosis* Ancient DNA Research

Spigelman and Lemma (65) were the first to demonstrate MTBC DNA in ancient skeletal material. The following year, it was detected in 1,000-year-old human tissue from an Andean mummy and confirmed by sequencing (66). This showed that tuberculosis was definitely present in the Americas before historical European contact. Thereafter, there were several reports of individual cases (6) and cases with no signs of paleopathology,

such as those in China from 2,000 years ago (67). Multiple burials enable populations to be studied and the epidemiology of past infections to be investigated. It is especially useful to study infections in the absence of any effective treatment because this has the potential to investigate the host–pathogen interaction at a molecular genetic level. In Thebes-West, ancient Egypt, tuberculosis was quite frequent across a long time period, from the Predynastic Period (c. 3500 to 2650 BC) to the Late Period (c. 1450 to 500 BC). It was suggested that the relatively high incidence of disease might have been related to the dense crowding in the city at a time of prosperity (68). Spoligotyping of the MTBC aDNA demonstrated human *M. tuberculosis* that had experienced the TbD1 deletion, similar to one of the major clades in the world today.

The earliest known published human cases of tuberculosis were from the Pre-Pottery Neolithic site of Atlit Yam in the eastern

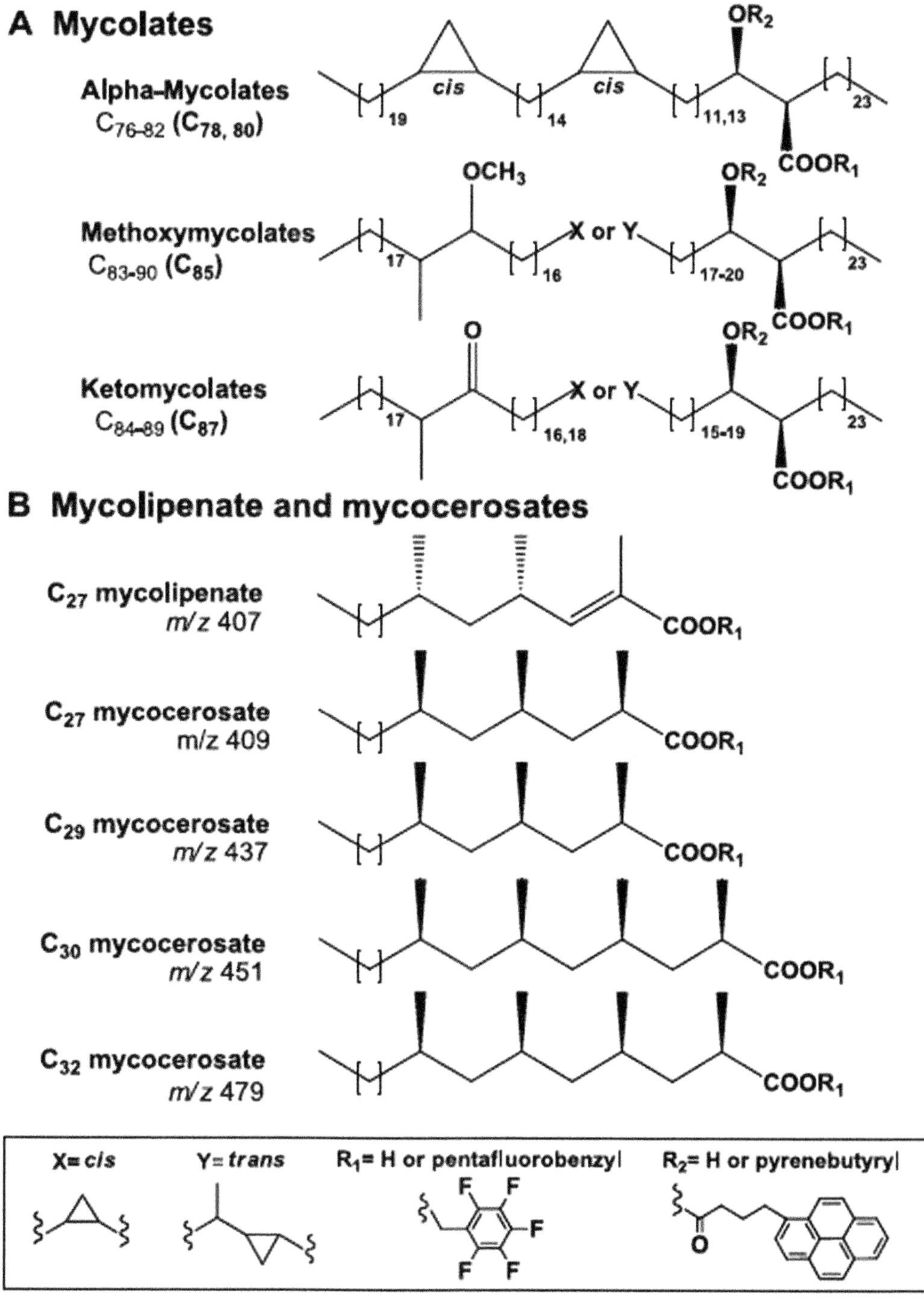

A Mycolates

B Mycolipenate and mycocerosates

FIGURE 4 Structures of *Mycobacterium tuberculosis* selected lipid biomarkers. (A) The main components of each mycolic acid class are shown; each class comprises a limited range of homologous components with different chain lengths. (B) Mycolipenic and mycocerosic acids; for each component, the ions (m/z) monitored on negative ion-chemical ionization gas chromatography-mass spectrometry (NICI-GCMS) of pentafluorobenzyl esters of these acids are given. (Courtesy of David E. Minnikin, School of Biosciences, University of Birmingham, Edgbaston, Birmingham, UK.)

Mediterranean, dating from 9250 to 8150 BP (16, 60). DNA preservation was excellent because the skeletal remains had been buried in thick clay under the sea. It was possible to demonstrate that two individuals were infected with a strain of *M. tuberculosis* in which the TbD1 deletion had occurred, thus identifying it as the human and not the bovine strain of the MTBC. In northern Europe, *M. tuberculosis* aDNA was detected in eight of 21 early Neolithic samples (5400 to 4800 BC) from central Germany (69), including three individuals with no visible pathology. Six samples were positive for spoligotyping. A further example of Late Neolithic tuberculosis was reported from central Hungary and dated to 7000 BP (70). In addition to molecular biomarkers, this was a striking case of tuberculosis with characteristic paleopathology—namely, hypertrophic pulmonary osteopathy with rib changes and cavitations in the vertebral bodies.

High-throughput sequencing of entire genomes, coupled with continuing updates in bioinformatics analysis, is now being applied to the examination of historical tuberculosis cases. Bouwman et al. (71) used next-generation sequencing, based on hybridization capture directed at specific polymorphic regions of the *M. tuberculosis* genome, to identify a detailed genotype for a historical *M. tuberculosis* strain from an individual buried in the 19th century in St. George's Crypt, Leeds, West Yorkshire, England. A recent high-profile study (23) examined skeletal material from a large number of pre- and post-contact sites in the New World. Samples were processed via established protocols and screened for MTBC DNA by an in-solution capture assay designed for the *rpoB*, *gyrA*, *gyrB*, *katG*, and *mpt40* genes. Capture products for samples and negative controls were sequenced on an Illumina MiSeq System and mapped to the corresponding regions in the *M. tuberculosis* H37Rv reference genome. There were three positive samples that had been recovered from excavations in Peru and derived from Chiribaya cultures associated

with the Middle Horizon/Late Intermediate Period (750 to 1350 AD).

Using a metagenomic approach of shotgun sequencing without prior enrichment, Chan et al. (72) identified two *M. tuberculosis* genomes in one 18th century naturally mummified individual from Vác, Hungary. The Vác mummies are remarkably well preserved because of the local environmental conditions in the sealed crypt where they were found. In addition, there is a contemporaneous archive, so family groups and age at death can be determined (73). The latest findings confirmed an earlier PCR-based study (74) in which each member of a small family group appeared to be infected with a different strain of *M. tuberculosis*. Whole-genome sequencing showed that the mother and her older daughter were both infected with the same two strains of *M. tuberculosis*, but in different proportions (75). In addition, of six other individuals in the same crypt, one was co-infected with three different strains of *M. tuberculosis*, two individuals were co-infected with two strains and the remaining three individuals were each infected with one strain. Six different sub-lineages were detected in this population. PCR-based genotyping has also demonstrated at least one possible mixed infection from British historical samples (76). The presence of mixed infections with more than one strain of *M. tuberculosis* is of particular interest because this phenomenon has been noted in modern tuberculosis infections and described as microevolution, both within a patient and between patients (77). Finding evidence of the same phenomenon in the pre-antibiotic era indicates that this phenomenon is related more to human population density than to antimicrobial therapy.

Human Infections with Other Members of the *Mycobacterium tuberculosis* Complex

In their spoligotyping study of three different populations in ancient Egypt, Zink et al. (68)

showed evidence of human *M. tuberculosis* that had experienced the TbD1 deletion. In addition, there were some strains lacking spacer 39, in samples from a Middle Kingdom tomb in Thebes-West (2050 to 1650 BC). This latter pattern is typical of *Mycobacterium africanum*. *M. bovis* is very rare in the archaeological record. However, it was found in a group of Iron Age Siberian pastoralists (4th century BC to 4th century AD) who wintered in huts with their animals (78). The paleopathogenic lesions (79) were typical of tuberculosis, and the analysis of *M. bovis*–specific genetic markers confirmed the diagnosis. The most recent example of an archaeological human infection with an animal lineage of the MTBC is the study from Peru (23). These ancient strains were most closely related to those adapted to seals and sea lions, known as *Mycobacterium pinnipedii*.

Past Human Migrations and *Mycobacterium tuberculosis* Epidemiology

There is a striking parallel between human lineage and the corresponding *M. tuberculosis* lineage that is harbored, which apparently persists even if people relocate to other parts of the world. A recent example is shown by a study of a locally dominant *M. tuberculosis* genetic lineage currently circulating among aboriginal populations in Alberta, Saskatchewan, and Ontario, as well as among French Canadians in Quebec, Canada (80). Substantial contact between these human populations was limited to a specific historical era (1710 to 1870 AD), when individuals met to barter furs. Therefore, this study of *M. tuberculosis* provides independent evidence of past contact between distinct peoples.

Applications of *Mycobacterium tuberculosis* Cell Wall Lipid Markers

Initially, lipid biomarkers were used as an independent method of detecting pathogenic mycobacteria and verifying aDNA data. For example, Redman et al. (61) investigated a group of 49 individuals from the 1837 to 1936 Coimbra Identified Skeletal Collection (Portugal), half of whom had records giving tuberculosis as a cause of death. There was a 72% correlation of the detection of mycocerosate acid biomarkers with individuals who were listed as likely to have died of tuberculosis. Because there is no amplification in lipid analysis, the amount of specific lipid biomarkers can be quantified and used for comparative purposes. Another use is the examination of archaeological samples in which there is poor or no preservation of aDNA. Although samples may show signs of diagenesis, especially samples several thousand years old, the lipid biomarkers are significantly more stable than aDNA and provide independent evidence of infection.

Use of Other Biomarkers

Carbohydrate or protein antigens are molecules that can induce antibody production, so they can be used to detect infectious organisms. These are often more stable than nucleic acids, but even so, antigenic determinants in ancient tissues may be damaged or destroyed, which limits their use. Antibodies may also be detected in mummified tissues by using a method such as an enzyme-linked immunoelectrotransfer blot. Although the host produces antibodies in response to an infection, their direct detection is difficult because they are generally less stable than antigens. A rare example of a study based on the host response to infection is the work by Corthals et al. (81). These authors reported the first use of shotgun proteomics to detect the protein expression profile of buccal swabs and cloth samples from two 500-year-old Andean mummies. The profile of one of the mummies was consistent with an immune system response to a severe bacterial lung infection at the time of death. One buccal swab contained a probable pathogenic *Mycobacterium* species that was confirmed by DNA amplification, sequencing, and phylogenetic analyses. However, the species was not determined.

An initial proteomic study using shotgun proteomics of mummified lung tissue from Vác, Hungary, revealed a suite of proteins, predominantly derived from the human host. Only one sample demonstrated weak evidence of organisms from the MTBC (82). It appears that most identified proteins were derived from high abundance human extracellular matrix proteins, although some immune system and catabolic proteins were identified.

Host Susceptibility and Ancient Tuberculosis

Co-infections

It is rare to find visual paleopathological changes that indicate more than one infection. However, in addition to specific aDNA markers, both *M. tuberculosis* and *Mycobacterium leprae* lipids can be identified and distinguished from each other. This led to the discovery that in the past some individuals were co-infected (83). In Europe, the decline of leprosy in the late middle Ages coincided with a rise in tuberculosis. Suggested reasons for this observation include cross-immunity (5) and the increased virulence of tuberculosis (83). Both scenarios are epidemiologically feasible (84).

Evidence of parasitic infections is widespread in human remains. Co-infection with *M. tuberculosis* and parasites is an important public health problem today in areas of the world where both are endemic and is therefore likely to have been so in the past. This has been demonstrated in ancient Lower Egypt dating to c. 800 BC, where four mummies were found with aDNA from both *M. tuberculosis* and *Plasmodium falciparum* (85). Chagas disease, caused by the protozoan parasite *Trypanosoma cruzi*, was prevalent in pre-Columbian northern Peru, and tuberculosis has also been demonstrated in this population (86). A combination of paleopathology and aDNA analysis demon-

strated both diseases in a 12-year-old girl from 910 to 935 BP (87). Leishmaniasis is caused by a protozoan flagellated parasite with a sand fly vector that is associated with acacia trees. Northern Sudan is a region where this disease is endemic today, and *Leishmania* kinetoplast aDNA has been detected from Early Christian Nubia and Middle Kingdom ancient Egypt, where the lack of acacia trees and sand flies led to the assumption that the infection had been spread by trade connections with Nubia (88). Based on aDNA analysis, tuberculosis and leishmaniasis co-infections have been confirmed in Early Christian Nubia (89). It is known that intestinal unicellular parasites and worms are responsible for immunomodulatory effects in their host (90), including the modulation of responses to tuberculosis infections (91). As intestinal parasites are often found in the remains of early human populations, it is highly likely that such modulation of the host response occurred in the past.

Co-morbidities

Natural resistance to infection is reduced by physical and mental stress, which in turn is caused by invasion, warfare, displacement, and exclusion from society due to stigma. Also, a pre-existing infection decreases innate host resistance and increases susceptibility to further infections. However, another important consideration is host genetic susceptibility to infectious diseases. Innate immunity is an important arm of the host antimycobacterial defenses that sense various pathogenic microbes by pattern recognition receptors. Toll-like receptors (TLRs) play a crucial role in the recognition of *M. tuberculosis* and other pathogenic mycobacteria (92). Host immune activation occurs only in the presence of functional TLRs. Therefore, any coding changes in TLRs are associated with a substantial drop in susceptibility to these pathogens.

Different types of cancer (neoplasms) have a detrimental effect on host resistance.

For example, Langerhans cell histiocytosis, now recognized as a neoplasm, has distinct paleopathology and has been diagnosed in skeletal remains. Langerhans cell histiocytosis is related to immune dysfunction, an increased risk for acquired infections, and early death. A case of archaeological Langerhans cell histiocytosis in an infant, who also had aDNA evidence of tuberculosis infection, has been recognized in one of the Vác Hungarian mummies (93). It is likely that the genetic impairment in the host immune response in such cases increases susceptibility to tuberculosis. In this same 18th century Hungarian population, a 37-year-old woman, with a massive vertebral deformity that would have reduced lung function and therefore increased susceptibility to infection, was found to have tuberculosis (94).

CONCLUDING REMARKS

The study of ancient tuberculosis based on aDNA, published in 1993, was the first to directly investigate a human infectious disease by using microbial aDNA. Since that date, the field has become recognized around the world, and an increasing range of microbial pathogens is being examined. We now have a clearer understanding of the occurrence of tuberculosis in the past, its epidemiology, and its geographical location. The paleomicrobiology of tuberculosis has verified historical records of past infections and confirmed or refuted the findings of paleopathologists, anthropologists, and archaeologists. Palaeomicrobiology enables the recognition of co-infections, multiple infections, and co-morbidities such as tuberculosis and cancer. Links with medical anthropology and biomedical archaeology enable data on human diet, society, location, migrations, stress, and trauma to be considered in relation to past tuberculosis infection and host susceptibility. Collaboration with geneticists and evolutionary biologists has increased our understanding of the origins of the MTBC,

M. tuberculosis, and the time scale for their emergence.

The newer technologies of high-throughput sequencing, bioinformatics, and metagenomics have made it possible to obtain a complete picture of the host and the microbial contents of samples based on skeletal or mummified remains. An unexpected finding was the discovery of mixed infections with different *M. tuberculosis* lineages. Initially, it was believed that these were linked to scenarios such as the one in modern sub-Saharan Africa, where there are highly dense human populations, many immunocompromised patients, and rising levels of antibiotic resistance. However, we now know that a high incidence of infection, with multiple strains of *M. tuberculosis*, occurred in 18th century Hungary, during a time of peace but also of a rising human population and the start of industrialization. In the present day, there is widespread human mobility around the world, huge changes in lifestyle, and evolutionary changes increasing exponentially in line with the human population. In this scenario, we need to know the origins and development of human microbial pathogens such as *M. tuberculosis* in order to better understand the future. Paleomicrobiology is one of the tools that we can use.

CITATION

Donoghue HD. 2016. Paleomicrobiology of human tuberculosis. Microbiol Spectrum 4(4):PoH-0003-2014.

REFERENCES

1. **WHO.** 2015. *Global Tuberculosis Report 2015.* World Health Organization, Geneva, Switzerland. http://www.who.int/tb/publications/global_report/en/.
2. **WHO.** 2016. *Latent Tuberculosis Infection.* World Health Organization, Geneva, Switzerland. http://www.who.int/tb/challenges/ltbi/en.
3. **Ortner DJ, Putschar WJG.** 1985. *Identification of Pathological Conditions in Human Skeletal Remains.* Smithsonian Institution Press, Washington, DC.

4. **Manchester K.** 1984. Tuberculosis and leprosy in antiquity: an interpretation. *Med Hist* **28:**162–173. http://dx.doi.org/10.1017/S0025727300035705.

5. **Aufderheide A, Rodriguez Martin C.** 1998. *The Cambridge Encyclopedia of Human Paleopathology.* Cambridge University Press, Cambridge, UK.

6. **Donoghue HD, Spigelman M, Greenblatt CL, Lev-Maor G, Bar-Gal GK, Matheson C, Vernon K, Nerlich AG, Zink AR.** 2004. Tuberculosis: from prehistory to Robert Koch, as revealed by ancient DNA. *Lancet Infect Dis* **4:**584–592. http://dx.doi.org/10.1016/S1473-3099(04)01133-8.

7. **Donoghue HD.** 2011. Insights gained from palaeomicrobiology into ancient and modern tuberculosis. *Clin Microbiol Infect* **17:**821–829. http://dx.doi.org/10.1111/j.1469-0691.2011.03554.x.

8. **Kelley MA, Micozzi MS.** 1984. Rib lesions in chronic pulmonary tuberculosis. *Am J Phys Anthropol* **65:**381–386. http://dx.doi.org/10.1002/ajpa.1330650407.

9. **Roberts CA, Boylston A, Buckley L, Chamberlain AC, Murphy EM.** 1998. Rib lesions and tuberculosis: the palaeopathological evidence. *Tuber Lung Dis* **79:**55–60. http://dx.doi.org/10.1054/tuld.1998.0005.

10. **Santos AL, Roberts CA.** 2006. Anatomy of a serial killer: differential diagnosis of tuberculosis based on rib lesions of adult individuals from the Coimbra Identified Skeletal Collection, Portugal. *Am J Phys Anthropol* **130:**38–49. http://dx.doi.org/10.1002/ajpa.20160.

11. **Hershkovitz I, Greenwald CM, Latimer B, Jellema LM, Wish-Baratz S, Eshed V, Dutour O, Rothschild BM.** 2002. Serpens endocrania symmetrica (SES): a new term and a possible clue for identifying intrathoracic disease in skeletal populations. *Am J Phys Anthropol* **118:**201–216. http://dx.doi.org/10.1002/ajpa.10077.

12. **Mays S, Taylor GM.** 2002. Osteological and biomolecular study of two possible cases of hypertrophic osteoarthropathy from Mediaeval England. *J Archaeol Sci* **29:**1267–1276. http://dx.doi.org/10.1006/jasc.2001.0769.

13. **Crubézy É, Ludes B, Poveda J-D, Clayton J, Crouau-Roy B, Montagnon D.** 1998. Identification of *Mycobacterium* DNA in an Egyptian Pott's disease of 5 400 years old. *C R Acad Sci Paris Sci de la Vie* **321:**941–951.

14. **Zink A, Haas CJ, Reischl U, Szeimies U, Nerlich AG.** 2001. Molecular analysis of skeletal tuberculosis in an ancient Egyptian population. *J Med Microbiol* **50:**355–366. http://dx.doi.org/10.1099/0022-1317-50-4-355.

15. **Formicola V, Milanesi Q, Scarsini C.** 1987. Evidence of spinal tuberculosis at the beginning of the fourth millennium BC from Arene Candide cave (Liguria, Italy). *Am J Phys Anthropol* **72:**1–6. http://dx.doi.org/10.1002/ajpa.1330720102.

16. **Hershkovitz I, Donoghue HD, Minnikin DE, Besra GS, Lee O-Y, Gernaey AM, Galili E, Eshed V, Greenblatt CL, Lemma E, Bar-Gal GK, Spigelman M.** 2008. Detection and molecular characterization of 9,000-year-old *Mycobacterium tuberculosis* from a Neolithic settlement in the Eastern Mediterranean. *PLoS One* **3:**e3426. http://dx.doi.org/10.1371/journal.pone.0003426.

17. **Tayles N, Buckley HR.** 2004. Leprosy and tuberculosis in Iron Age Southeast Asia? *Am J Phys Anthropol* **125:**239–256. http://dx.doi.org/10.1002/ajpa.10378.

18. **Suzuki T, Fujita H, Choi JG.** 2008. Brief communication: new evidence of tuberculosis from prehistoric Korea-Population movement and early evidence of tuberculosis in far East Asia. *Am J Phys Anthropol* **136:**357–360. http://dx.doi.org/10.1002/ajpa.20811.

19. **Allison MJ, Mendoza D, Pezzia A.** 1973. Documentation of a case of tuberculosis in Pre-Columbian America. *Am Rev Respir Dis* **107:**985–991.

20. **Gerszten E, Allison MJ, Maguire B.** 2012. Paleopathology in South American mummies: a review and new findings. *Pathobiology* **79:**247–256. http://dx.doi.org/10.1159/000334087.

21. **Arrieta MA, Bordach MA, Mendonça OJ.** 2014. Pre-Columbian tuberculosis in northwest Argentina: skeletal evidence from Rincón Chico 21 cemetery. *Int J Osteoarchaeol* **24:**1–14. http://dx.doi.org/10.1002/oa.1300.

22. **Arriaza BT, Salo W, Aufderheide AC, Holcomb TA.** 1995. Pre-Columbian tuberculosis in northern Chile: molecular and skeletal evidence. *Am J Phys Anthropol* **98:**37–45. http://dx.doi.org/10.1002/ajpa.1330980104.

23. **Bos KI, Harkins KM, Herbig A, Coscolla M, Weber N, Comas I, Forrest SA, Bryant JM, Harris SR, Schuenemann VJ, Campbell TJ, Majander K, Wilbur AK, Guichon RA, Wolfe Steadman DL, Cook DC, Niemann S, Behr MA, Zumarraga M, Bastida R, Huson D, Nieselt K, Young D, Parkhill J, Buikstra JE, Gagneux S, Stone AC, Krause J.** 2014. Pre-Columbian mycobacterial genomes reveal seals as a source of New World human tuberculosis. *Nature* **514:**494–497. http://dx.doi.org/10.1038/nature13591.

24. **Blaser MJ, Kirschner D.** 2007. The equilibria that allow bacterial persistence in human hosts. *Nature* **449:**843–849. http://dx.doi.org/10.1038/nature06198.

25. **Armelagos GJ, Goodman AH, Jacobs KH.** 1991. The origins of agriculture: population growth during a period of declining health. *Popul Environ* **13:**9–22. http://dx.doi.org/10.1007/BF01256568.

26. **McMichael AJ.** 2001. Human culture, ecological change, and infectious disease: are we experiencing history's fourth great transition? *Ecosyst Health* **7:**107–115. http://dx.doi.org/10.1046/j.1526-0992.2001.007002107.x.

27. **Barnes I, Duda A, Pybus OG, Thomas MG.** 2011. Ancient urbanization predicts genetic resistance to tuberculosis. *Evolution* **65:**842–848. http://dx.doi.org/10.1111/j.1558-5646.2010.01132.x.

28. **Armelagos GJ, Brown PJ, Turner B.** 2005. Evolutionary, historical and political economic perspectives on health and disease. *Soc Sci Med* **61:**755–765. http://dx.doi.org/10.1016/j.socscimed.2004.08.066.

29. **Donoghue HD.** 2009. Human tuberculosis—an ancient disease, as elucidated by ancient microbial biomolecules. *Microbes Infect* **11:**1156–1162. http://dx.doi.org/10.1016/j.micinf.2009.08.008.

30. **Clark GA, Kelley MA, Grange JM, Hill MC, Katzenberg MA, Klepinger LL, Lovell NC, McGrath J, Micozzi MS, Steinbock RT.** 1987. The evolution of mycobacterial disease in human populations: a reevaluation [and comments and reply]. *Curr Anthropol* **28:**45–62. http://dx.doi.org/10.1086/203490.

31. **Brosch R, Gordon SV, Marmiesse M, Brodin P, Buchrieser C, Eiglmeier K, Garnier T, Gutierrez C, Hewinson G, Kremer K, Parsons LM, Pym AS, Samper S, van Soolingen D, Cole ST.** 2002. A new evolutionary scenario for the *Mycobacterium tuberculosis* complex. *Proc Natl Acad Sci U S A* **99:**3684–3689. http://dx.doi.org/10.1073/pnas.052548299.

32. **Smith NH, Hewinson RG, Kremer K, Brosch R, Gordon SV.** 2009. Myths and misconceptions: the origin and evolution of *Mycobacterium tuberculosis*. *Nat Rev Microbiol* **7:**537–544. http://dx.doi.org/10.1038/nrmicro2165.

33. **WHO.** 2014. Xpert MTB/RIF implementation manual. Geneva, Switzerland. ISBN 9789241506700.

34. **Pääbo S, Poinar H, Serre D, Jaenicke-Després V, Hebler J, Rohland N, Kuch M, Krause J, Vigilant L, Hofreiter M.** 2004. Genetic analyses from ancient DNA. *Annu Rev Genet* **38:**645–679. http://dx.doi.org/10.1146/annurev.genet.37.110801.143214.

35. **O'Rourke DH, Hayes MG, Carlyle SW.** 2000. Ancient DNA studies in physical anthropology. *Annu Rev Anthropol* **29:**217–242. http://dx.doi.org/10.1146/annurev.anthro.29.1.217.

36. **Höss M, Jaruga P, Zastawny TH, Dizdaroglu M, Pääbo S.** 1996. DNA damage and DNA sequence retrieval from ancient tissues. *Nucleic Acids Res* **24:**1304–1307. http://dx.doi.org/10.1093/nar/24.7.1304.

37. **Poinar HN, Hofreiter M, Spaulding WG, Martin PS, Stankiewicz BA, Bland H, Evershed RP, Possnert G, Pääbo S.** 1998. Molecular coproscopy: dung and diet of the extinct ground sloth *Nothrotheriops shastensis*. *Science* **281:**402–406. http://dx.doi.org/10.1126/science.281.5375.402.

38. **Poinar HN.** 2003. The top 10 list: criteria of authenticity for DNA from ancient and forensic samples. *Int Congr Ser* **1239:**573–579. http://dx.doi.org/10.1016/S0531-5131(02)00624-6.

39. **Nguyen-Hieu T, Aboudharam G, Drancourt M.** 2012. Heat degradation of eukaryotic and bacterial DNA: an experimental model for paleomicrobiology. *BMC Res Notes* **5:**528. http://dx.doi.org/10.1186/1756-0500-5-528.

40. **Minnikin DE, Lee OY-C, Wu HHT, Besra GS, Donoghue HD.** 2012. Molecular biomarkers for ancient tuberculosis, p 1–36. *In* Cardona P-J (ed), *Understanding Tuberculosis – Deciphering the Secret Life of the Bacilli*. InTech Open Access Publisher, Rijeka, Croatia. http://www.intechopen.com/books/understanding-tuberculosis-deciphering-the-secret-life-of-the-bacilli/molecular-biomarkers-for-ancient-tuberculosis

41. **Dziadek J, Sajduda A, Boruń TM.** 2001. Specificity of insertion sequence-based PCR assays for *Mycobacterium tuberculosis* complex. *Int J Tuberc Lung Dis* **5:**569–574.

42. **Tanaka MM, Small PM, Salamon H, Feldman MW.** 2000. The dynamics of repeated elements: applications to the epidemiology of tuberculosis. *Proc Natl Acad Sci U S A* **97:**3532–3537. http://dx.doi.org/10.1073/pnas.97.7.3532.

43. **Taylor GM, Stewart GR, Cooke M, Chaplin S, Ladva S, Kirkup J, Palmer S, Young DB.** 2003. Koch's bacillus - a look at the first isolate of *Mycobacterium tuberculosis* from a modern perspective. *Microbiology* **149:**3213–3220. http://dx.doi.org/10.1099/mic.0.26654-0.

44. **Ottoni C, Koon HEC, Collins MJ, Penkman KEH, Rickards O, Craig OE.** 2009. Preservation of ancient DNA in thermally damaged archaeological bone. *Naturwissenschaften* **96:**267–278. http://dx.doi.org/10.1007/s00114-008-0478-5.

45. **Witt N, Rodger G, Vandesompele J, Benes V, Zumla A, Rook GA, Huggett JF.** 2009. An assessment of air as a source of DNA contamination encountered when performing PCR. *J Biomol Tech* **20:**236–240.

46. **Taylor GM, Mays SA, Huggett JF.** 2010. Ancient DNA (aDNA) studies of man and microbes: general similarities, specific differences. *Int J Osteoarchaeol* **20:**747–751. http://dx.doi.org/10.1002/oa.1077.

47. **Kamerbeek J, Schouls L, Kolk A, van Agterveld M, van Soolingen D, Kuijper S, Bunschoten A, Molhuizen H, Shaw R, Goyal M, van Embden J.** 1997. Simultaneous detection and strain differentiation of *Mycobacterium tuberculosis* for diagnosis and epidemiology. *J Clin Microbiol* **35:**907–914.

48. **Gutierrez MC, Brisse S, Brosch R, Fabre M, Omaïs B, Marmiesse M, Supply P, Vincent V.** 2005. Ancient origin and gene mosaicism of the progenitor of *Mycobacterium tuberculosis*. *PLoS Pathog* **1:**e5. http://dx.doi.org/10.1371/journal.ppat.0010005.

49. **Hirsh AE, Tsolaki AG, DeRiemer K, Feldman MW, Small PM.** 2004. Stable association between strains of *Mycobacterium tuberculosis* and their human host populations. *Proc Natl Acad Sci U S A* **101:**4871–4876. http://dx.doi.org/10.1073/pnas.0305627101.

50. **Reed MB, Pichler VK, McIntosh F, Mattia A, Fallow A, Masala S, Domenech P, Zwerling A, Thibert L, Menzies D, Schwartzman K, Behr MA.** 2009. Major *Mycobacterium tuberculosis* lineages associate with patient country of origin. *J Clin Microbiol* **47:**1119–1128. http://dx.doi.org/10.1128/JCM.02142-08.

51. **Firdessa R, Berg S, Hailu E, Schelling E, Gumi B, Erenso G, Gadisa E, Kiros T, Habtamu M, Hussein J, Zinsstag J, Robertson BD, Ameni G, Lohan AJ, Loftus B, Comas I, Gagneux S, Tschopp R, Yamuah L, Hewinson G, Gordon SV, Young DB, Aseffa A.** 2013. Mycobacterial lineages causing pulmonary and extrapulmonary tuberculosis, Ethiopia. *Emerg Infect Dis* **19:**460–463. http://dx.doi.org/10.3201/eid1903.120256.

52. **Yen S, Bower JE, Freeman JT, Basu I, O'Toole RF.** 2013. Phylogenetic lineages of tuberculosis isolates in New Zealand and their association with patient demographics. *Int J Tuberc Lung Dis* **17:**892–897. http://dx.doi.org/10.5588/ijtld.12.0795.

53. **Boritsch EC, Supply P, Honoré N, Seemann T, Stinear TP, Brosch R.** 2014. A glimpse into the past and predictions for the future: the molecular evolution of the tuberculosis agent. *Mol Microbiol* **93:**835–852. http://dx.doi.org/10.1111/mmi.12720.

54. **Supply P, Marceau M, Mangenot S, Roche D, Rouanet C, Khanna V, Majlessi L, Criscuolo A, Tap J, Pawlik A, Fiette L, Orgeur M, Fabre M, Parmentier C, Frigui W, Simeone R, Boritsch EC, Debrie AS, Willery E, Walker D, Quail MA, Ma L, Bouchier C, Salvignol G, Sayes F, Cascioferro A, Seemann T, Barbe V, Locht C, Gutierrez MC, Leclerc C, Bentley SD, Stinear TP, Brisse S, Médigue C, Parkhill J, Cruveiller S, Brosch R.** 2013. Genomic analysis of smooth tubercle bacilli provides insights into ancestry and pathoadaptation of *Mycobacterium tuberculosis*. *Nat Genet* **45:**172–179. http://dx.doi.org/10.1038/ng.2517.

55. **Wirth T, Hildebrand F, Allix-Béguec C, Wölbeling F, Kubica T, Kremer K, van Soolingen D, Rüsch-Gerdes S, Locht C, Brisse S, Meyer A, Supply P, Niemann S.** 2008. Origin, spread and demography of the *Mycobacterium tuberculosis* complex. *PLoS Pathog* **4:**e1000160. http://dx.doi.org/10.1371/journal.ppat.1000160.

56. **Comas I, Coscolla M, Luo T, Borrell S, Holt KE, Kato-Maeda M, Parkhill J, Malla B, Berg S, Thwaites G, Yeboah-Manu D, Bothamley G, Mei J, Wei L, Bentley S, Harris SR, Niemann S, Diel R, Aseffa A, Gao Q, Young D, Gagneux S.** 2013. Out-of-Africa migration and Neolithic coexpansion of *Mycobacterium tuberculosis* with modern humans. *Nat Genet* **45:**1176–1182. http://dx.doi.org/10.1038/ng.2744.

57. **Minnikin DE, Lee OY-C, Wu HHT, Nataraj V, Donoghue HD, Ridell M, Watanabe M, Alderwick L, Bhatt A, Besra GS.** 2015. Pathophysiological implications of cell envelope structure in *Mycobacterium tuberculosis* and related taxa. *In* Ribón W (ed), *Tuberculosis*. InTech Open Access Publisher, Rijeka, Croatia. http://www.intechopen.com/books/tuberculosis-expanding-knowledge/pathophysiological-implications-of-cell-envelope-structure-in-mycobacterium-tuberculosis-and-related. http://dx.doi.org/10.5772/59585.

58. **Donoghue HD, Spigelman M, Zias J, Gernaey-Child AM, Minnikin DE.** 1998. *Mycobacterium tuberculosis* complex DNA in calcified pleura from remains 1400 years old. *Lett Appl Microbiol* **27:**265–269. http://dx.doi.org/10.1046/j.1472-765X.1998.00436.x.

59. **Gernaey AM, Minnikin DE, Copley MS, Dixon RA, Middleton JC, Roberts CA.** 2001. Mycolic acids and ancient DNA confirm an osteological diagnosis of tuberculosis. *Tuberculosis (Edinb)* **81:**259–265. http://dx.doi.org/10.1054/tube.2001.0295.

60. **Lee OY-C, Wu HHT, Donoghue HD, Spigelman M, Greenblatt CL, Bull ID, Rothschild BM, Martin LD, Minnikin DE, Besra GS.** 2012. *Mycobacterium tuberculosis* complex lipid virulence factors preserved in the 17,000-year-old skeleton of an extinct bison, *Bison antiquus*. *PLoS One* **7:**e41923. http://dx.doi.org/10.1371/journal.pone.0041923.

61. **Redman JE, Shaw MJ, Mallet AI, Santos AL, Roberts CA, Gernaey AM, Minnikin DE.** 2009. Mycocerosic acid biomarkers for the diagnosis of tuberculosis in the Coimbra Skeletal Collection. *Tuberculosis (Edinb)* **89:**267–277. http://dx.doi.org/10.1016/j.tube.2009.04.001.

62. **Lee OY-C, Bull ID, Molnár E, Marcsik A, Pálfi G, Donoghue HD, Besra GS, Minnikin DE.** 2012. Integrated strategies for the use of lipid biomarkers in the diagnosis of ancient mycobacterial disease, p 63–69. *In* Mitchell PD, Buckberry J (ed), *Proceedings of the Twelfth Annual Conference of the British Association for Biological Anthropology and Osteoarchaeology, Department of Archaeology and Anthropology University of Cambridge 2010.* BAR International Series 2380, Archaeopress, Oxford, UK.

63. **Lee OY-C, Wu HHT, Besra GS, Rothschild BM, Spigelman M, Hershkovitz I, Bar-Gal GK, Donoghue HD, Minnikin DE.** 2015. Lipid biomarkers provide evolutionary signposts for the oldest known cases of tuberculosis. *Tuberculosis (Edinb)* **95(Suppl 1):**S127–S132 http://dx.doi.org/10.1016/j.tube.2015.02.013.

64. **Minnikin DE, Besra GS, Lee OY-C, Spigelman M, Donoghue HD.** 2011. The interplay of DNA and lipid biomarkers in the detection of tuberculosis and leprosy, p 109–114. *In* Gill-Frerking G, Rosendahl W, Zink A, Piombino-Mascali D (ed), *Yearbook of Mummy Studies 1.* Verlag Dr. Friedrich Pfeil, Münich, Germany.

65. **Spigelman M, Lemma E.** 1993. The use of the polymerase chain reaction (PCR) to detect *Mycobacterium tuberculosis* in ancient skeletons. *Int J Osteoarchaeol* **3:**137–143. http://dx.doi.org/10.1002/oa.1390030211.

66. **Salo WL, Aufderheide AC, Buikstra J, Holcomb TA.** 1994. Identification of *Mycobacterium tuberculosis* DNA in a pre-Columbian Peruvian mummy. *Proc Natl Acad Sci U S A* **91:**2091–2094. http://dx.doi.org/10.1073/pnas.91.6.2091.

67. **Fusegawa H, Wang B-H, Sakurai K, Nagasawa K, Okauchi M, Nagakura K.** 2003. Outbreak of tuberculosis in a 2000-year-old Chinese population. *Kansenshogaku Zasshi* **77:**146–149. http://dx.doi.org/10.11150/kansenshogakuzasshi1970.77.146.

68. **Zink AR, Molnár E, Motamedi N, Pálfy G, Marcsik A, Nerlich AG.** 2007. Molecular history of tuberculosis from ancient mummies and skeletons. *Int J Osteoarchaeol* **17:**380–391. http://dx.doi.org/10.1002/oa.909.

69. **Nicklisch N, Maixner F, Ganslmeier R, Friederich S, Dresely V, Meller H, Zink A, Alt KW.** 2012. Rib lesions in skeletons from early neolithic sites in Central Germany: on the trail of tuberculosis at the onset of agriculture. *Am J Phys Anthropol* **149:**391–404. http://dx.doi.org/10.1002/ajpa.22137.

70. **Masson M, Molnár E, Donoghue HD, Besra GS, Minnikin DE, Wu HHT, Lee OY-C, Bull ID, Pálfi G.** 2013. Osteological and biomolecular evidence of a 7000-year-old case of hypertrophic pulmonary osteopathy secondary to tuberculosis from neolithic hungary. *PLoS One* **8:**e78252. http://dx.doi.org/10.1371/journal.pone.0078252.

71. **Bouwman AS, Kennedy SL, Müller R, Stephens RH, Holst M, Caffell AC, Roberts CA, Brown TA.** 2012. Genotype of a historic strain of *Mycobacterium tuberculosis. Proc Natl Acad Sci U S A* **109:**18511–18516. http://dx.doi.org/10.1073/pnas.1209444109.

72. **Chan JZ-M, Sergeant MJ, Lee OY-C, Minnikin DE, Besra GS, Pap I, Spigelman M, Donoghue HD, Pallen MJ.** 2013. Metagenomic analysis of tuberculosis in a mummy. *N Engl J Med* **369:**289–290. http://dx.doi.org/10.1056/NEJMc1302295.

73. **Fletcher HA, Donoghue HD, Holton J, Pap I, Spigelman M.** 2003. Widespread occurrence of *Mycobacterium tuberculosis* DNA from 18th-19th century Hungarians. *Am J Phys Anthropol* **120:**144–152. http://dx.doi.org/10.1002/ajpa.10114.

74. **Fletcher HA, Donoghue HD, Taylor GM, van der Zanden AGM, Spigelman M.** 2003. Molecular analysis of *Mycobacterium tuberculosis* DNA from a family of 18th century Hungarians. *Microbiology* **149:**143–151. http://dx.doi.org/10.1099/mic.0.25961-0.

75. **Kay GL, Sergeant MJ, Zhou Z, Chan Z-M, Millard A, Quick J, Szokossy I, Pap I, Spigelman M, Loman NJ, Achtman M, Donoghue HD, Pallen MJ.** 2015. Eighteenth-century genomes show that mixed infections were common at time of peak tuberculosis in Europe. *Nature Communications* **6:**6717. http://dx.doi.org/10.1038/ncomms7717.

76. **Müller R, Roberts CA, Brown TA.** 2014. Genotyping of ancient *Mycobacterium tuberculosis* strains reveals historic genetic diversity. *Proc Biol Sci* **281:**20133236. http://dx.doi.org/10.1098/rspb.2013.3236.

77. **Pérez-Lago L, Comas I, Navarro Y, González-Candelas F, Herranz M, Bouza E, García-de-Viedma D.** 2014. Whole genome sequencing analysis of intrapatient microevolution in *Mycobacterium tuberculosis*: potential impact on the inference of tuberculosis transmission. *J Infect Dis* **209:**98–108. http://dx.doi.org/10.1093/infdis/jit439.

78. **Taylor GM, Murphy E, Hopkins R, Rutland P, Chistov Y.** 2007. First report of *Mycobacterium bovis* DNA in human remains from the

Iron Age. *Microbiology* **153**:1243–1249. http://dx.doi.org/10.1099/mic.0.2006/002154-0.

79. **Murphy EM, Chistov YK, Hopkins R, Rutland P, Taylor GM.** 2009. Tuberculosis among Iron Age individuals from Tyva, South Siberia: palaeopathological and biomolecular findings. *J Archaeol Sci* **36**:2029–2038. http://dx.doi.org/10.1016/j.jas.2009.05.025.

80. **Pepperell CS, Granka JM, Alexander DC, Behr MA, Chui L, Gordon J, Guthrie JL, Jamieson FB, Langlois-Klassen D, Long R, Nguyen D, Wobeser W, Feldman MW.** 2011. Dispersal of *Mycobacterium tuberculosis* via the Canadian fur trade. *Proc Natl Acad Sci U S A* **108**:6526–6531. http://dx.doi.org/10.1073/pnas.1016708108.

81. **Corthals A, Koller A, Martin DW, Rieger R, Chen EI, Bernaski M, Recagno G, Dávalos LM.** 2012. Detecting the immune system response of a 500 year-old Inca mummy. *PLoS One* **7**:e41244. http://dx.doi.org/10.1371/journal.pone.0041244.

82. **Hendy J, Collins M, Teoh KY, Ashford DA, Thomas-Oates J, Donoghue HD, Pap I, Minnikin DE, Spigelman M, Buckley M.** 2016. The challenge of identifying tuberculosis proteins in archaeological tissues. *J Archaeol Sci* **66**:146–153. http://dx.doi.org/10.1016/j.jas.2016.01.003.

83. **Donoghue HD, Marcsik A, Matheson C, Vernon K, Nuorala E, Molto JE, Greenblatt CL, Spigelman M.** 2005. Co-infection of *Mycobacterium tuberculosis* and *Mycobacterium leprae* in human archaeological samples: a possible explanation for the historical decline of leprosy. *Proc Biol Sci* **272**:389–394. http://dx.doi.org/10.1098/rspb.2004.2966.

84. **Hohmann N, Voss-Böhme A.** 2013. The epidemiological consequences of leprosy-tuberculosis co-infection. *Math Biosci* **241**:225–237. http://dx.doi.org/10.1016/j.mbs.2012.11.008.

85. **Lalremruata A, Ball M, Bianucci R, Welte B, Nerlich AG, Kun JFJ, Pusch CM.** 2013. Molecular identification of falciparum malaria and human tuberculosis co-infections in mummies from the Fayum depression (Lower Egypt). *PLoS One* **8**:e60307. http://dx.doi.org/10.1371/journal.pone.0060307.

86. **Aufderheide AC, Salo W, Madden M, Streitz J, Buikstra J, Guhl F, Arriaza B, Renier C, Wittmers LE Jr, Fornaciari G, Allison M.** 2004. A 9,000-year record of Chagas' disease. *Proc Natl Acad Sci U S A* **101**:2034–2039. http://dx.doi.org/10.1073/pnas.0307312101.

87. **Arrieza BT, Cartmell LL, Moragas C, Nerlich AG, Salo W, Madden M, Aufderheide AC.** 2008. The bioarchaeological value of human mummies without provenience. *Chungara (Arica)* **40**:55–65.

88. **Zink AR, Spigelman M, Schraut B, Greenblatt CL, Nerlich AG, Donoghue HD.** 2006. Leishmaniasis in ancient Egypt and Upper nubia. *Emerg Infect Dis* **12**:1616–1617. http://dx.doi.org/10.3201/eid1210.060169.

89. **Spigelman M, Greenblatt CL, Vernon K, Zylber MI, Sheridan SG, Van Gerven DP, Shaheem Z, Hansraj F, Donoghue HD.** 2005. Preliminary findings on the paleomicrobiological study of 400 naturally mummified human remains from upper Nubia. *J Biol Res* **80**:91–95. ISSN: 1826–8838.

90. **Fenton A.** 2013. Dances with worms: the ecological and evolutionary impacts of de-worming on coinfecting pathogens. *Parasitology* **140**:1119–1132. http://dx.doi.org/10.1017/S0031182013000590.

91. **Li X-X, Zhou X-N.** 2013. Co-infection of tuberculosis and parasitic diseases in humans: a systematic review. *Parasit Vectors* **6**:79. http://dx.doi.org/10.1186/1756-3305-6-79.

92. **Thada S, Valluri VL, Gaddam SL.** 2013. Influence of Toll-like receptor gene polymorphisms to tuberculosis susceptibility in humans. *Scand J Immunol* **78**:221–229. http://dx.doi.org/10.1111/sji.12066.

93. **Spigelman M, Pap I, Donoghue HD.** 2006. A death from Langerhans cell histiocytosis and tuberculosis in 18th century Hungary - what palaeopathology can tell us today. *Leukemia* **20**:740–742. http://dx.doi.org/10.1038/sj.leu.2404103.

94. **Kustár Á, Pap I, Végvári Z, Kristóf LA, Pálfi G, Karlinger K, Kovács B, Szikossy I.** 2011. Use of 3D virtual reconstruction for pathological investigation and facial reconstruction of an 18th century mummified nun from Hungary, p 83–93. *In* Gill-Frerking G, Rosendahl W, Zink A, Piombino-Mascali D (ed), *Yearbook of Mummy Studies 1.* Verlag Dr. Friedrich Pfeil, Münich, Germany.

95. **Évinger S, Bernert Z, Fóthi E, Wolff K, Kővári I, Marcsik A, Donoghue HD, O'Grady J, Kiss KK, Hajdu T.** 2011. New skeletal tuberculosis cases in past populations from western Hungary (Transdanubia). *HOMO - J Comp Hum Biol* **62**:165–183. http://dx.doi.org/10.1016/j.jchb.2011.04.001.

96. **Hajdu T, Donoghue HD, Bernert Z, Fóthi E, Kővári I, Marcsik A.** 2012. A case of spinal tuberculosis from the middle ages in Transylvania (Romania). *Spine* **37**:E1598–E1601. http://dx.doi.org/10.1097/BRS.0b013e31827300dc.

97. **Galagan JE.** 2014. Genomic insights into tuberculosis. *Nat Rev Genet* **15**:307–320. http://dx.doi.org/10.1038/nrg3664.

Paleomicrobiology of Leprosy 13

MARK SPIGELMAN[1,2,3] and MAURO RUBINI[4,5]

Leprosy (or Hansen's disease) is a chronic infectious disease caused by a slowly multiplying obligate pathogen, *Mycobacterium leprae*, an acid-fast, rod-shaped bacillus belonging to a single species with limited genetic variability (1). *M. leprae* has four types and 16 subtypes based on single-nucleotide polymorphisms (SNPs) and variations (InDels) in insertions and deletions (2). Although not highly infectious, it is transmitted via droplets from the nose and mouth during close and frequent contacts with untreated cases. The incubation period can be very long (in some cases up to 20 years) before clinical signs and symptoms become apparent (3, 4). Leprosy can affect all age groups and both sexes. To this day, we cannot grow the bacillus in the laboratory. The bacillus is almost specific to humans but does affect some armadillos (*Dasypus novemcinctus*) from the southeastern United States (Texas and Louisiana) (5) and a specific type of monkey. In the laboratory, rats, mice, and hamsters can be infected. The diagnosis of the disease in archaeological specimens can be problematic because the characteristic bony changes can occur in a number of other diseases. Thus, paleomicrobiology

[1]Centre for Clinical Microbiology, Division of Infection & Immunity, University College London, London, UK; [2]Department of Anatomy and Anthropology, Sackler Faculty of Medicine, Tel Aviv University, Tel Aviv, Israel; [3]The Kuvin Center for the Study of Infectious and Tropical Diseases and Ancient DNA, Hadassah Medical School, The Hebrew University, Jerusalem, Israel; [4]Department of Archaeology Foggia University, Foggia, Italy; [5]Anthropological Service of Soprintendenza per i Beni Archeologici del Lazio (Ministry of Culture), Rome, Italy.

Paleomicrobiology of Humans
Edited by Michel Drancourt and Didier Raoult
© 2016 American Society for Microbiology, Washington, DC
doi:10.1128/microbiolspec.PoH-0009-2015

can help to confirm a clinical incidence and, along with genotyping, can trace the spread of the disease and even human migration patterns responsible for its spread (6).

Today, leprosy is curable, and multidrug therapy (MDT) that provides a simple yet highly effective cure for all types of leprosy has been made available by the WHO free of charge to all patients worldwide since 1995. The incidence of new cases, however, appears not to be decreasing according to official reports received from 115 countries; the global registered prevalence of leprosy at the end of 2012 was 189,018 cases. The number of new cases reported globally in 2014 was 249,657, compared with 232,857 in 2012. Pockets where the disease is highly endemic still remain in some areas of many countries, and statistics show that 220,810 new cases of leprosy (95%) were reported from 16 countries and only 5% of new cases from the rest of the world (7).

Today, the clinical approach to leprosy is multidisciplinary because the disease is of interest to specialists from various fields, including dermatology, infectious diseases, ophthalmology, neurology, microbiology, and molecular biology. Clinical signs of disease can include relatively painless ulcers, skin lesions consisting of hypopigmented macules (flat, pale areas of skin), and eye damage (dryness and reduction in the blinking reflex). However, leprosy mainly affects the peripheral nervous system, especially in the extremities (neural leprosy), with secondary involvement of the skin and other tissues. The Schwann cells, the principal support cells in the peripheral nervous system, appear to be the major target of *M. leprae* in peripheral nerves. As a consequence of long-term infection and the host immunological response, Schwann cells are ultimately destroyed or functionally impaired in damaged nerves. Atrophy of myelinated fibers and secondary demyelization have been shown to occur as result of infection, sometimes even in the absence of evidence of inflammation or acid-fast bacilli.

In an advanced stage, leprosy can affect the skeleton, and a number of specific and nonspecific bony and osteoporotic changes occur during pathogenesis (8; M. Rubini and P. Zaio, submitted for publication). Specific bone changes are the result of invasion of the tissues by *M. leprae*, which is why facial changes are seen especially in the nose, where nerves are in the mucosa attached to the bone. Secondary bone changes are a consequence of peripheral nerve involvement (hands, feet, and lower leg bones) and ultimately lead to sensory nerve anesthesia with trauma to the hands and feet because the victim cannot feel pain, resulting in ulceration of the feet and hands and then secondary infection of the bones of the hands and feet. Motor and autonomic nerve damage, causing lesions such as claw hands and feet, and bone absorption and remodeling are all features of advanced disease. Osteoporosis secondary to disuse results in a reduction of the bony mass and is caused by three factors: a reduction of osteoblastic activity, a rise of osteoclastic activity, or both (simultaneous competition). In leprosy, nerve damage and the consequent cessation of normal functional activity of the limb produce an imbalance in the osteoclastic–osteoblastic equilibrium in the bones, with a rise of osteoclastic activity (resorption of bone) that leads to osteoporosis by disuse (3). Osteoporosis can produce remodeling of the bone because of the collapse of the cortical layer of the bone under the pressure of the muscles. This type of damage causes remodeling and is responsible for the characteristic pencil shape of the bones of the fingers and toes (3, 9). However, the major classic foci of leprosy damage are in the rhinomaxillary region of the face and the tubular bones of the hands and feet, although other bones may be affected (10, 11).

The body response to the disease is highly variable. The immune status of each infected individual determines the type and severity of pathological changes (12). The different clinical manifestations are the result of the immune status of patients, not the bacterial variety (12). At one end of the immune spectrum, there is low resistance to infection or multibacillary (MeSH lepromatous) leprosy; at the other, there is high resistance to infection, resulting in paucibacillary (MeSH tuberculoid) leprosy (13). Between these extremes are the borderline types. However, such responses may help a clinician to place the patient in the more detailed Ridley-Jopling immune spectrum classification (14). In addition to these groups manifesting clinical leprosy, individuals may be infected with the bacteria but not develop clinical signs of the disease; this is termed *subclinical leprosy* (3). According to some authors (13, 15), the incidence of subclinical leprosy is as high today as it was in the past (16). In a general view of the disease, it is important also to consider the infectivity of the other forms of leprosy (15) because in ancient times these probably played an important role in the spread of the disease (16, 17).

BIOLOGICAL HISTORY OF LEPROSY

Leprosy was described in osteoarchaeological remains for the first time in India and dated to 2000 BCE (18). Molecular studies on SNP distribution have shown that the disease started following the migration paths of the first human groups from East Africa in the direction of Asia and established itself in eastern and central Europe and in the Mediterranean Basin about 40,000 years ago (1, 2). This result suggests that even the last Neanderthals and the first *Homo sapiens* came into contact with *M. leprae*. The question is, why we do not find hominids with the bone changes of leprosy? The answer is complex, with the main problem being the scarcity of hominid fossil remains showing changes in the rhinomaxillary region and tubular bones of the hand and foot. The second could be linked to the long incubation period and slow development of the disease. Leprosy involves the skeleton (only and not always) in the last stage of its clinical course. In this case, age at death could have played an important role in the possibility of finding of skeletal changes in hominid remains. Another possibility is that in the immune system of the Neanderthals, a balance of immunity developed between *M. leprae* and host, in which the effects caused by the mycobacterium were attenuated and better tolerated. Leprosy was preferentially a rural disease in the past, as it is today (16, 19). It is probably for this reason that in European prehistoric remains it is very rare to find leprosy cases. The most ancient European skeletal evidence may be the relatively recent discovery in Scotland of an individual with leprosy dating from 1600 to 2000 BC (20; J. Roberts, personal communication). Furthermore, aside from a case in northeastern Italy (21), we do not know of many other cases in the first millennium BCE in Europe. During the first millennium AD, the documented osteoarchaeological remains become significant. There are skeletal traces of leprosy during the Middle Ages in western Europe—for example, in Denmark (22, 23, 24, 25, 26), southern Germany (27), England (28, 29), Ireland (30), France (28), and Italy (31, 32, 33, 34), whereas in central Europe, there are cases in the Czech Republic (35) and Hungary (36). The presence of a case of lepromatous leprosy in central Asia (Uzbekistan) dating to the 1st to 3rd centuries AD (37) suggests that during the first millennium, the spread of leprosy was primarily terrestrial (1, 9). The disease was known in the Roman Empire, as will be detailed later in this chapter.

DIAGNOSIS OF LEPROSY IN PALEOPATHOLOGY

Leprosy afflicts bones at the late stages of disease, and only a small percentage of them. The typical changes can be best appreciated by looking at the figures (Fig. 1, 2, 3, 4). The bone changes are highly variable in intensity and location. Even though

Stage of bone infiltration by *Mycobacterium leprae*

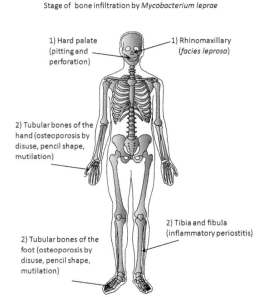

1) Hard palate (pitting and perforation)

1) Rhinomaxillary (*facies leprosa*)

2) Tubular bones of the hand (osteoporosis by disuse, pencil shape, mutilation)

2) Tibia and fibula (inflammatory periostitis)

2) Tubular bones of the foot (osteoporosis by disuse, pencil shape, mutilation)

FIGURE 1 Stages of bone infiltration.

the first infiltration of the mycobacterium involves the peripheral nerves of the limbs, the first region of the skeleton that is infiltrated is the rhinomaxillary region because here the nerves are in close contact with bone and covered by a thin layer of nasal mucosa. The so-called facies leprosa, or rhinomaxillary syndrome (38), appears when the skeletal structure of the nose is remodeled and capped with new bone and the nasal spine disappears. In this stage, remodeling of the margins of the nasal aperture in conjunction with alterations to the intranasal structures results in the appearance of a wide, empty cavity. Regression of the inferior portion of the superior face is clearly visible. Alterations of the tubular bones of the hands and feet (especially the fifth metatarsal bone, where the same nerves that innervate the fibula are present) are secondary to pathological changes in the peripheral nerves, with consequent loss of sensory and motor functions (39). This leads to osteoporosis due to disuse, with consequences in the metacarpal and metatarsal

bones producing the characteristic shape of a pencil tip (Fig. 5).

Although the pathogenesis of the skeletal changes in leprosy is generally difficult to assess in osteoarchaeological remains, Ortner (10) observed that the presence of destructive and proliferative lesions could indicate the infectious phase (chronic or acute) of the disease, without specifying the pathological process that caused the bone changes. Evidence of a roughened, finely pitted cortical surface is uncommon in osteoarchaeological samples. Ortner suggests that the destructive phase is characterized by brief osteoclastic activity quickly followed by slight repair that smoothes the surface. Because in a number of pathological conditions there may be a considerable destruction of bone in the rhinomaxillary region, the differential diagnosis must take account of these conditions; fungal infections such as aspergillosis and mucormycosis (phycomycosys), actinomycosis (a bacterial rather than a true fungal disease), lupus vulgaris (tuberculosis of the facial skin and soft tissue), granulomatous diseases such as sarcoidosis, and treponemal diseases such as syphilis can

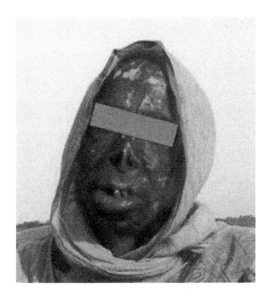

FIGURE 2 Facial appearance of a victim.

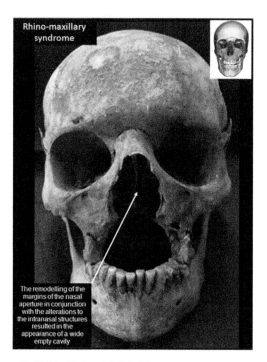

Rhino-maxillary syndrome

The remodelling of the margins of the nasal aperture in conjunction with the alterations to the intranasal structures resulted in the appearance of a wide empty cavity

Skull Sk19 from Winchester (UK). Roffey S, Tucker K . International Journal of Paleopathology 2 (2012) 170-180. Reprinted with permission of the publisher.

FIGURE 3 Rhinomaxillary syndrome.

cause similar changes in the rhinomaxillary region (10, 40). Psoriatic arthritis, septic arthritis, and other joint diseases may also cause identical changes in the hands and feet.

PALEOMICROBIOLOGY

This new branch of science was developed following the discovery of PCR by Mullis and Faloona (41), which enables the amplification of minute quantities of DNA and so makes it possible to sequence the results. The first paper on this topic was by Spigelman and Lemma (42) and was followed soon after by the paper of Salo et al. (43), who used a primer specific for *Mycobacterium tuberculosis* based on the insertion sequence found by Eisenach et al. (44). This amplified a specific sequence of human *M. tuberculosis* DNA that is only 123 bp in length. This was

an important development because as a result of the degradation of DNA over time, it is necessary to use such short target sequences of DNA from ancient bones in order to obtain positive results.

The search for *M. leprae* DNA began with a metatarsal bone from a grave at the site of a massacre of Christians by the Persians in 614 AD (45). It was excavated from the Monastery of Saint John the Baptist on the Jordan River, at the spot where it is traditionally believed that John the Baptist baptized Jesus. This is also the traditional site for the washing of the "leper" in Christian sources. The bone proved positive for *M. leprae* ancient DNA (aDNA) (46, 47). The set of primers and the technique developed

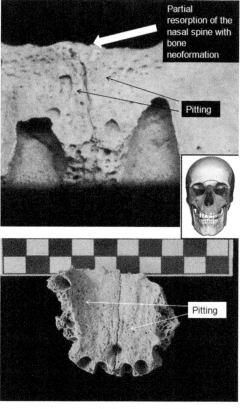

Partial resorption of the nasal spine with bone neoformation

Pitting

Pitting

A case of leprosy from Palombara (Rome, Italy, 5th century CE)

FIGURE 4 Nasal changes in leprosy.

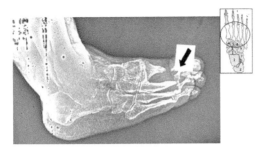

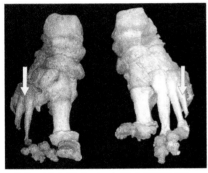

A - X-ray right foot. Recent case of lepromatous leprosy. Note the characteristic pencil shape of the tubular bones of the foot

B – Leper's feet. Note the characteristic pencil shape of the tubular bones (metatarsal) of the feet. Science Museum of London.

FIGURE 5 (A, B) Characteristic pencil shape of the tubular (metatarsal) bones of the foot.

by Hartskeerl et al. (48) was used to amplify a nucleotide sequence of 530 bp of the gene encoding the 36-kDa antigen of *M. leprae*, and a positive result was obtained. The 530-bp target sequence is a very large size for aDNA, but it was successful from this and subsequently from other samples after stringent precautions had been taken to avoid contamination. Therefore, this was deemed an early sign of the good preservation that can be found in *M. leprae*. It is believed that the lipid-rich cell wall of the organism was a significant contributing factor in this preservation.

A comprehensive review of the development of *M. leprae* paleomicrobiology was given in 1999 at a meeting in Bradford, UK ("The Past and Present of Leprosy") and subsequently published (49, 50). The size of the primers was considered a disadvantage

in the examination of archaeological samples, in which the DNA is likely to be damaged and fragmented. Two new sets of *M. leprae*-specific nested primers were designed. These were based on the 18-kDa antigen, which gave an outer product of 136 bp and inner product of 110 bp, as well as primers based on the RLEP repetitive sequence, which yielded a 129-bp outer product and a 99-bp nested product. The 18-kDa protein outer primers were 100-fold more sensitive than those for the 36-kDa antigen, and the RLEP outer primers were 1000-fold more sensitive (51). Subsequently, by using the above primers, a problem of the differential diagnosis of Madura foot and leprosy was resolved, in part by showing that the lesions proved positive for the DNA of *M. leprae*; however, this did not totally exclude Madura foot because of the possibility of the presence of fungal co-infection, thus demonstrating the limits of the conclusion one can draw even when positive results are obtained (52, 53, 54, 55).

The problems associated with the detection of aDNA by PCR are that amplification requires strict protocols to prevent contamination. This led to the development of a strict set of standards by workers studying human and mammalian aDNA, summarized in a paper by Cooper and Poinar (56). One recommendation was having independent duplication of data, but this was at times difficult. Bacterial DNA is not ubiquitous and can be highly localized. Thus, specimens sent to an independent laboratory may prove negative, and at the same time, multiple sampling of bones may damage important paleopathological specimens. In part, this problem was overcome by the identification of species-specific mycolic acids for *M. leprae* in the cell wall of the bacillus that allowed independent confirmation of the presence of the microbes. Because of the exquisite sensitivity of this technique, amplification with all its attendant problems is not necessary. Further, mycolic acids are chemically more robust then DNA and have been detected in specimens in which the DNA was not found (57, 58).

Following the sequencing of the entire *M. leprae* genome and the recognition of informative SNPs, scientists were able to determine the molecular typing of the organism (1). It was shown that the genome is remarkably stable over time, having only a human host (3). There is a clonal relationship between *M. leprae* and its human host, so discovering the genetic profiles of modern and extinct or ancient strains of *M. leprae* can tell us about the migration and spread of pathogen and host over time (1, 59, 60, 61). As technology progressed, the development of high-throughput sequencing enabled the sequencing of entire genomes from archaeological samples. The problem of low yields of aDNA was solved by next-generation sequencing, which uses targeted enrichment techniques with hybridization capture, followed by the use of bioinformatics software to construct the whole genome. Monot et al. (2) were able to conduct comparative genomic and phylogeographical analyses of *M. leprae*. A more recent paper reported a genome-wide comparison between medieval and modern *M. leprae* specimens (61). Furthermore, insights into ancient leprosy and tuberculosis gained by using metagenomics, as reported in papers Taylor et al (62) and Donoghue (63), demonstrated the progress of the science.

The use of paleomicrobiology in answering questions related to questions raised by archaeologists in leprosy can best be shown by a few examples. An early-published case based on aDNA resulted from the discovery of the Firsst-Century Tomb of the Shroud in Alkadema, Jerusalem. A body was found that had been buried in a wall niche, whereas all the other burials in this tomb were reburials after decomposition in an ossuary, as was the custom at the time. This raised the question, why? There was no convincing evidence of leprosy in the skeletal remains, but it was considered that perhaps there was something about the person that had made the family wary of touching the remains. The possibility of leprosy was suggested, and subsequent tests found the aDNA of *M. leprae* in the bones of this individual (64). Another question regarded the appearance and subsequent disappearance of the almost plaguelike epidemic of leprosy in medieval Europe. There is no question that the disease was present during the Roman Empire and even before (21, 65, 66). However, the skeletal evidence indicates that these sporadic early reports were not in the same epidemic proportions as in the subsequent medieval epidemic that swept through Europe but then almost disappeared over time. It has been suggested that the decline of leprosy in the Western World in the past was not monocausal but due to a complex web of social, legal, political, and biomedical causes (67). The latter is explained by the rise in tuberculosis. Studies that began in the 20th century proposed a model of cross-immunity between these two mycobacterial infections as a possible cause of the decline of leprosy. Lowe and McNulty (68), following the Rogers hypothesis (1924) and the overall work of Chaussinand (69), observed that the Bacille Calmette-Guérin (BCG) vaccine, prepared specifically for inoculation against tuberculosis, was of benefit in preventing leprosy in some people. The origin of this cross-immunity is controversial, but one fact is certain: in clinical contexts, leprosy may be prevented following BCG vaccination in 20% to 91% of individuals, based on different studies (70). There is a close antigenic relationship between the mycobacteria causing leprosy (*M. leprae*) and those causing tuberculosis (*M. tuberculosis*). The amino acid sequences of 65-kDa antigens of *M. leprae*, *M. tuberculosis*, and *Mycobacterium bovis* BCG display greater than 95% homology (71, 72). The higher reproductive rate of tubercle bacilli, compared with that of leprosy bacilli, and their degree of cross-immunity do not normally allow both infections to occur simultaneously, but there have been sporadic reports of the co-existence of tuberculosis and leprosy in the same patients based on physical symptoms (73, 74). Furthermore, a study based on the examination of mycobacterial aDNA by PCR has also investigated the relationship between tuberculosis and leprosy among individuals buried in the past, finding evidence for the co-existence of tuberculosis and leprosy (64, 74). This supports an earlier report that tuberculosis can occur in people throughout the immune spectrum of leprosy (75).

Chaussinand (69) was the first to observe that in several different clinical settings, the prevalence of leprosy was inversely related to that of tuberculosis. Therefore, it may be considered unusual to encounter their co-existence. For example, an important epidemiological study from South Africa reported an increased incidence of pulmonary tuberculosis among patients with leprosy, but not vice versa (76). This theory of cross-immunity between tuberculosis and leprosy led to it being proposed as the hypothesis for the disappearance of leprosy from western Europe in the 14th century AD, before the advent of chemotherapy (16, 77). However, research on skeletal samples, albeit limited, has more recently refuted the hypothesis that this cross-immunity exists and has suggested an alternative explanation for the decline of leprosy in western Europe (74). Furthermore, Manchester (16) and Donoghue et al. (74) suggested that immunological changes seen in multibacillary leprosy (lepromatous/low resistance), together with the socioeconomic impact of the disease, may have led to increased mortality due to tuberculosis that resulted in the decline of leprosy. Despite this long history of epidemiological, clinical, and microbiological studies, the exact relationship between tuberculosis and leprosy still remains unclear. The only certain fact is that there is homology between the two mycobacteria, and that in Europe during the Middle Ages an increase of tuberculosis as a result of increasing urbanization co-incided with a decline in leprosy (78). Paleopathologists have noted the presence of leprosy in Roman times and before, but not in the almost plague-level proportions found in medieval Europe, when almost every town has its own leprosarium. Indeed, Manchester and Roberts (79) noted that in Britain the number of lazar houses for the care of leprosy patients increased dramatically during the 12th and early 13th centuries. One possible suggestion, supported by aDNA and cell wall lipid analysis, is that the migration of tribes from central Asia during the 1st millennium led to a separate introduction of leprosy into eastern and central Europe (80).

CONCLUSION

It can be seen that the use of paleomicrobiological techniques in the study of leprosy has the potential to assist clinicians and paleopathologists in many important aspects of their work. How can it help clinicians? In leprosy, because of the unique nature of the organism, paleomicrobiological techniques can help solve problems of differential diagnosis. In the case of co-infection with *Mycobacterium tuberculosis*, they can also suggest a cause of death and possibly even trace the migratory patterns of people in antiquity.

CITATION

Spigelman M, Rubini M. 2016. Paleomicrobiology of leprosy, Microbiol Spectrum 4(4): PoH-0009-2015.

REFERENCES

1. **Monot M, Honoré N, Garnier T, Araoz R, Coppée J-Y, Lacroix C, Sow S, Spencer JS, Truman RW, Williams DL, Gelber R, Virmond M, Flageul B, Cho SN, Ji B, Paniz-Mondolfi A, Convit J, Young S, Fine PE, Rasolofo V, Brennan PJ, Cole ST.** 2005. On the origin of leprosy. *Science* **308:**1040–1042.
2. **Monot M, Honoré N, Garnier T, Zidane N, Sherafi D, Paniz-Mondolfi A, Matsuoka M, Taylor GM, Donoghue HD, Bouwman A, Mays S, Watson C, Lockwood D, Khamispour A, Dowlati Y, Jianping S, Rea TH, Vera-Cabrera L, Stefani MM, Banu S, Macdonald M, Sapkota BR, Spencer JS, Thomas J, Harshman K, Singh P, Busso P, Gattiker A, Rougemont J, Brennan**

PJ, Cole ST. 2009. Comparative genomic and phylogeographic analysis of *Mycobacterium leprae*. *Nat Genet* **41**:1282–1288.

3. **Robbins SL, Cotran RS.** 2002. *Pathologic Basis of Disease*. W.B. Saunders Co., Philadelphia, PA.

4. **Jopling WH.** 2004. *Handbook of Leprosy*, 5th ed. Sheridan Medical Books, New York, NY.

5. **Maiden MCJ.** 2009. Putting leprosy on the map. *Nat Genet* **41**(12):1264.

6. **Rubini M, Zaio P, Roberts AC.** 2014. Tuberculosis and leprosy in Italy. New skeletal evidence. *HOMO - J Comp Hum Biol* **65**:13–32.

7. **WHO.** 2016. *Leprosy fact sheet*. http://www.who.int/mediacentre/factsheets/fs101/en.

8. **Thappa DM, Sharma VK, Kaur S, Suri S.** 1992. Radiological changes in hands and feet in disabled leprosy patients: a clinical-radiological correlation. *Lepr India* **64**:58–66.

9. **Kulkarni VN, Mehta JM.** 1983. Tarsal disintegration (T.D.) in leprosy. *Lepr India* **55**:338–370.

10. **Ortner DJ.** 2003. *Identification of Pathological Conditions in Human Skeletal Remains*. Smithsonian Institution Press, Washington, DC.

11. **Roberts CA, Manchester K.** 2005. *The archaeology of disease*, 3rd ed. Sutton Publishing, Stroud, Gloucestershire, UK.

12. **Geluk A, Ottenhoff THM.** 2006. HLA and leprosy in the pre- and postgenomic eras. *Hum Immunol* **67**:439–445.

13. **Kumarasinghe SPW, Kumarasinghe MP, Amarasinghe UTP.** 2005. "Tap sign" in tubercoloid and borderline tuberculoid leprosy. *Int J Lepr* **72**:291–295.

14. **Ridley DS, Jopling WH.** 1966. Classification of leprosy according to immunity, a five-group system. *Int J Lepr* **34**:255–273.

15. **Kampirapap K.** 2008. Assessment of subclinical leprosy infection through the measurement of PGL-1 antibody levels in residents of a former leprosy colony in Thailand. *Lepr Rev* **79**:315–319.

16. **Manchester K.** 1984. Tuberculosis and leprosy in antiquity: an interpretation. *Med Hist* **28**:162–173.

17. **Rubini M, Zaio P.** 2009. Lepromatous leprosy in an early mediaeval cemetery in Central Italy (Morrione, Campochiaro, Molise, 6th–7th century AD). *J Archeol Sci* **36**:2771–2779.

18. **Robbins G, Mushrif Tipathy V, Misra VN, Mohanty RK, Shinde VS, Gray KM, Schug MD.** 2009. Ancient skeletal evidence for leprosy in India (2000 B.C.). *Plos One* **4**:1–8. doi:10.1371/journal.pone.0005669.

19. **Kerr-Pontes LR, Dorta Montenegro AC, Lima Barreto M, Werneck GL, Feldmeier H.** 2004. Inequality and leprosy in Northeast Brazil: an ecological study. *Int J Ecol* **33**:262–269.

20. **Anonymous.** 2002. Bones raise leprosy doubts. *BBC News*, November 5. http://news.bbc.co.uk/2/hi/uk_news/scotland/2406001.stm.

21. **Mariotti V, Dutour O, Belcastro MG, Facchini F, Brasili P.** 2005. Probable early presence of leprosy in Europe in a Celtic skeleton of the 4th-3rd century BC (Casalecchio di Reno, Bologna, Italy). *Int J Osteoarchaeol* **15**:311–325.

22. **Andersen JG.** 1969. *Studies on the Medieval Diagnosis of Leprosy in Denmark. An Osteoarchaeological, Historical and Clinical Study*. Costers Bogtrykkeri, Copenhagen, Denmark.

23. **Arcini C.** 1999. Health and disease in early Lund: osteopathological studies of 3,305 individuals buried in the first cemetery area of Lund 990-1536. *Archaeologica Lundensia VIII*. Department of Community Health Sciences, Lund.

24. **Boldsen JL.** 2001. Epidemiological approach to the paleopathological diagnosis of leprosy. *Am J Phys Anthropol* **115**:380–387.

25. **Boldsen JL.** 2005. Leprosy and mortality in the medieval Danish village of Tirup. *Am J Phys Anthropol* **126**:159–168.

26. **Bennike P, Lewis ME, Schultkowski H, Valentin F.** 2005. Comparison of child morbidity in two contrasting Medieval cemeteries from Denmark. *Am J Phys Anthropol* **128**:734–746.

27. **Haas CJ, Zink A, Szeimies U, Nerlich G.** 2002. Molecular evidence of *Mycobacterium leprae* in historical bone samples from South Germany, p 287–292. *In* Roberts CA, Lewis ME, Manchester K (ed), *The Past and Present of Leprosy. Archaeological, Historical, Palaeopathological and Clinical Approaches*. British Archaeological Reports International Series (Book 1054). British Archaeological Reports, Oxford, UK.

28. **Roberts CA.** 2000. Infectious disease in biocultural perspective: past, present and future work in Britain, p 145–162. *In* Cox M, Mays S (ed), *Human Osteology in Archaeology and Forensic Science*. Greenwich Medical Media, London, UK.

29. **Schultz M, Roberts C.** 2002. Diagnosis of leprosy in skeletons from an English late Medieval hospital using histological analysis, p 89–105. *In* Roberts CA, Lewis ME, Manchester K (ed), *The Past and Present of Leprosy. Archaeological, Historical, Palaeopathological and Clinical Approaches*. British Archaeological Reports International Series (Book 1054). British Archaeological Reports, Oxford, UK.

30. **Murphy E, Manchester K.** 2002. Evidence for leprosy in Medieval Ireland, p 193–199. *In* Roberts CA, Lewis ME, Manchester K (ed), *The Past and Present of Leprosy. Archaeological, Historical, Palaeopathological and Clinical Approaches.* British Archaeological Reports International Series (Book 1054). British Archaeological Reports, Oxford, UK.

31. **Belcastro MG, Mariotti V, Facchini F, Dutour O.** 2005. Leprosy in a skeleton from the 7th century Necropolis of Vicenne-Campochiaro (Molise, Italy). *Int J Osteoarchaeol* 15:431–448.

32. **Rubini M.** 2008. Disfiguring diseases and society in the fourth - fifth centuries A.D.: the case of Palombara Sabina (Rome, Central Italy). *In* Proceedings of the Annual Meeting Archaeological Institute of America, Chicago, IL.

33. **Rubini M, Dell'Anno V, Giuliani R, Favia P, Zaio P.** 2012. The first probable case of leprosy in Southeast Italy (13th -14th centuries AD, Montecorvino, Puglia). *J Anthropol.* doi:10.1155/2012/262790.

34. **Rubini M, Cerroni V, Zaio P.** 2015. *The Origin and Spread of* Mycobacterium leprae *in Human Evolution.* Konrad Lorenz Institute, Vienna, Austria.

35. **Strouhal E, Ladislava H, Likovsky J, Vargova L, Danes J.** 2002. Traces of leprosy from the Czech Kingdom, p 223–232. *In* Roberts CA, Lewis ME, Manchester K (ed), *The Past and Present of Leprosy. Archaeological, Historical, Palaeopathological and Clinical Approaches.* British Archaeological Reports International Series (Book 1054). British Archaeological Reports, Oxford, UK.

36. **Palfi G.** 1991. The first osteoarchaeological evidence of leprosy in Hungary. *Int J Osteoarchaeol* 1:99–102.

37. **Blau S, Yagodin V.** 2005. Osteoarchaeological evidence for leprosy from Western Central Asia. *Am J Phys Anthropol* 126:150–158.

38. **Andersen JG, Manchester K.** 1992. The rhinomaxillary syndrome in leprosy: a clinical, radiological and paleopathological study. *Int J Osteoarchaeol* 2:121–129.

39. **Bullock WE.** 1990. Mycobacterium leprae (leprosy), p 1906–1914. *In* Mandell GL, Douglas RG, Bennett JE (ed), *Principles and Practice of Infectious Diseases.* Churchill Livingstone, New York, NY.

40. **Job CK, Sushil M, Chandi SM.** 2001. *Differential Diagnosis of Leprosy: a Guide Book for Histopathologists.* Sasakawa Memorial Health Foundation, Tokyo, Japan.

41. **Mullis K, Faloona F.** 1987. Specific synthesis of DNA in vitro via a polymerase catalysed chain reaction. *Methods Enzymol* 155:335–350.

42. **Spigelman M, Lemma E.** 1992. The use of the polymerase chain reaction to detect *Mycobacterium tuberculosis* in ancient skeletons. *Int J Osteoarchaeol* 3:137–143.

43. **Salo W, Aufterheide A, Buikstra J, Holcomb T.** 1994. Identification of *Mycobacterium tuberculosis* DNA in a pre-Columbian Peruvian mummy. *Proc Natl Acad Sci U S A* 91:2091–2094.

44. **Eisenach KD, Cave MD, Bates JH, Crawford JT.** 1990. Polymerase chain reaction amplification of a repetitive DNA sequence specific for *Mycobacterium tuberculosis. J Infect Dis* 161:977–981.

45. **Gill M.** 1992. *A History of Palestine 634-1099.* Cambridge University Press, Cambridge, UK.

46. **Rafi A, Spigelman M, Stanford J, Lemma E, Donoghue H, Zias J.** 1994. *Mycobacterium leprae* from ancient bones detected by PCR. *Lancet* 343:1360–1361.

47. **Rafi A, Spigelman M, Stanford J, Lemma E, Donoghue H, Zias J.** 1994. DNA of *Mycobacterium leprae* detected by PCR in ancient bone. *Int J Osteoarchaeol* 4:287–290.

48. **Hartskeerl RA, De Wit M, Yand Klaster PR.** 1989. Polymerase chain reaction for the detection of *Mycobacterium leprae. J Gen Microbiol* 135:2355–2364.

49. **Donoghue HD, Gladykowska-Rzeczycka J, Marcsik A, Holton J, Spigelman M.** 2002. *Mycobacterium leprae* in archaeological samples, p 271–285. *In* Roberts CA, Lewis ME, Manchester K (ed), *The Past and Present of Leprosy: Archaeological, Historical, Palaeopathological and Clinical Approaches.* British Archaeological Reports Series. Archaeopress, Oxford, UK.

50. **Spigelman M, Donoghue HD.** 2002. The study of ancient DNA answers a palaeopathological question, p 287–296. *In* Roberts CA, Lewis ME, Manchester K (ed), *The Past and Present of Leprosy: Archaeological, Historical, Palaeopathological and Clinical Approaches.* British Archaeological Reports Series. Archaeopress, Oxford, UK.

51. **Donoghue HD, Holton J, Spigelman M.** 2001. PCR primers that can detect low levels of *Mycobacterium leprae* DNA. *J Med Microbiol* 50:177–182.

52. **Hershkovitz I, Spiers M, Katznelson A, Arensburg B.** 1992. Unusual pathological condition in the lower extremities of a skeleton from ancient Israel. *Am J Phys Anthropol* 88:23–26.

53. **Hershkovitz I, Spiers M, Arensburg B.** 1993. Leprosy or Madura foot? The ambiguous nature of infectious disease in paleopathology:

reply to Dr Manchester. *Am J Phys Anthropol* **91**:251–253.

54. **Manchester K.** 1993. Unusual pathological condition in the lower extremities of a skeleton from ancient Israel. *Am J Phys Anthropol* **91**:249–250.

55. **Spigelman M, Donoghue HD.** 2001. Brief communication: unusual pathological condition in the lower extremities of a skeleton from ancient Israel. *Am J Phys Anthropol* **114**:92–93.

56. **Cooper A, Poinar HN.** 2000. Ancient DNA: do it right or not at all. *Science* **289**:1139.

57. **Lee OY-C, Bull ID, Molnár E, Marcsik A, Pálfi G, Donoghue HD, Besra GS, Minnikin DE.** 2012. Integrated strategies for the use of lipid biomarkers in the diagnosis of ancient mycobacterial disease, p 63–69. *In* Mitchell PD, Buckberry J (ed), *Proceedings of the Twelfth Annual Conference of the British Association for Biological Anthropology and Osteoarchaeology, Department of Archaeology and Anthropology, University of Cambridge 2010*. British Arachaeological Reports International Series (Book 2380). Archaeopress, Oxford, UK.

58. **Minnikin DE, Besra GS, Lee O-YC, Spigelman M, Donoghue HD.** 2011. The interplay of NA and lipid biomarkers in the detection of tuberculosis and leprosy in mummies and other skeletal remains, p 109–114. *In* Gill-Frerking H, Rosendahl W, Zink A, Piombini-Mascali D. *Yearbook of Mummy Studies*, **vol. 1**. Verlag Dr. Friedrich Pfeil, München, Germany.

59. **Watson CL, Popescu E, Boldsen J, Slaus M, Lockwood DNJ.** 2010. Correction: single nucleotide polymorphism analysis of European archaeological *M. leprae* DNA. *PLoS One* **5**(1). doi:10.1371/annotation/1b400b6e-8883-436c-b3c4-00e1ec2db101.

60. **Economou C, Kjellström A, Lidén K, Panagopoulos I.** 2013. Ancient DNA reveals an Asian type of *Mycobacterium leprae* in medieval Scandinavia. *J Archaeol Sci* **40**:465–470.

61. **Shuenemann VJ, Singh P, Mendum TA, Krause-Kyora B, Jäger G, Bos KI, Herbig A, Economou C, Benjak A, Busso P.** 2013. Genome-wide comparison of medieval and modern *Mycobacterium leprae*. *Science* **341**:179–183.

62. **Taylor GM, Tucker K, Butler R, Pike AWG, Lewis J, Roffey PM, Lee OY-C, Wu HHT, Minnikin DE, Besra GS, Singh P, Cole ST, Stewart GR.** 2013. Detection and strain typing of ancient *Mycobacterium leprae* from a medieval leprosy hospital. *PLoS One* **8**:e62406. doi:10.1371/journal.pone.0062406.

63. **Donoghue HD.** 2013. Insights into ancient leprosy and tuberculosis using metagenomics. *Trends Microbiol* **21**:448–450.

64. **Matheson CD, Vernon KK, Lahti A, Fratpietro R, Spigelman M, Gibson S, Greenblatt CL, Donoghue HD.** 2010. Correction: Molecular exploration of the First-Century Tomb of the Shroud in Akeldama, Jerusalem. *PLoS One* **5**(4). doi:10.1371/annotation/32ada7b9-3772-4c08-9135-b5c0933f0b5e. Zissu, Boaz [added].

65. **Molto JE.** 2002. p 179–192. *In* Roberts CA, Lewes ME, Manchester K (ed), *The Past and Present of Leprosy. Archaeological, Historical, Palaeopathological and Clinical Approaches*. British Archaeological Research International Series (Book 1054). Archaeopress, Oxford, UK.

66. **Rubini M, Erdal YS, Spigelman M, Zaio P, Donoghue HD.** 2014. Paleopathological and molecular study on two cases of ancient childhood leprosy from the Roman and Byzantine empires. *Int J Osteoarchaeol* **24**:570–582.

67. **Rawcliffe C.** 2006. *Leprosy in Medieval England*. Boydell Press, Woodbridge, UK.

68. **Lowe J, McNulty F.** 1953. Tuberculosis and leprosy: immunological studies in healthy persons. *Br Med J* **ii**:579–584.

69. **Chaussinand R.** 1948. Tuberculose et lèpre, maladies antagoniques. Eviction de la lèpre parl la tuberculose. *Int J Lepr* **16**:431–438.

70. **Merle CSC, Cunha SS, Rodrigues LC.** 2010. BCG vaccination and leprosy protection: review of current evidence and status of BCG in leprosy control. *Expert Rev Vaccines* **9**:209–222.

71. **Shinnick TM, Sweetser D, Thole J.** 1987. The etiologic agents of leprosy and tuberculosis share an immunoreactive protein antigen with the vaccine strain *Mycobacterium bovis* BCG. *Infect Immun* **55**:1932–1935.

72. **Prasad R, Verma SK, Singh R.** 2010. Concomittant pulmonary tuberculosis and borderline leprosy with type-II lepra reaction in single patient. *Lung India* **27**:19–23.

73. **Ravindra K, Sugareddy M, Ramachander T.** 2010. Coexistence of borderline tuberculoid Hansen's disease with tuberculosis verrucosa cutis in a child – a rare case. *Lepr India* **82**:91–93.

74. **Donoghue HD, Marcsik A, Matheson C, Vernon K, Nuorala E, Molto JE, Greenblatt CL, Spigelman M.** 2005. Co-infection of *Mycobacterium tuberculosis* and *Mycobacterium leprae* in human archaeological samples: a possible explanation for the historical decline of leprosy. *Proc R Socy of Lond B Biol Sci* **272**:389–394.

75. **Kumar B, Kaur S, Kataria S, Roy SN.** 1982. Concomitant occurrence of leprosy and tuberculosis – a clinical, bacteriological and radiological evaluation. *Lepr India* **54:**71–76.

76. **Gartner EMS, Glatthaar E, Imkamp FMJH, Kok SH.** 1980. Association of tuberculosis and leprosy in South Africa. *Lepr Rev* **51:**5–10.

77. **Lietman T, Porco T, Blower S.** 1997. Leprosy and tuberculosis: the epidemiological consequences of cross-immunity. *Am J Public Health* **87:**1923–1927.

78. **Hohmann H, Voss-Böhme A.** 2013. The epidemiological consequences of leprosy-tuberculosis co-infection. *Math Biosci* **241:**225–237.

79. **Manchester K, Roberts C.** 1989. The palaeopathology of leprosy in Britain: a review. *World Archaeol* **21:**265–272.

80. **Donoghue G, Taylor M, Marcsik A, Molnár E, Pálfi G, Pap I, Teschler-Nicola M, Pinhasi R, Erdal YS, Velemínsky P, Likovsky J, Belcastro MG, Mariotti V, Riga A, Rubini M, Zaio P, Besra GS, Lee OYC, Wu HHT, Minnikin DE, Bull ID, O'Grady J, Spigelman M.** 2015. A migration-driven model for the historical spread of leprosy in medieval eastern and Central Europe. *Infect Genet Evol* **31:**250–256.

Past Intestinal Parasites

MATTHIEU LE BAILLY[1] and ADAUTO ARAÚJO[2,†]

The study of ancient parasites, paleoparasitology, is an area of parasitology that evolved as a research field combining archaeology, anthropology, biology, and health sciences (1). It aims to detect parasite traces in ancient samples and to study parasitism evolution over time and space (2). This research field covers the study of parasites in humans and other animal remains recovered from archaeological and paleontological sites.

Historically, the first record of an ancient parasite was revealed at the beginning of the 20th century by Sir Marc A. Ruffer, who discovered calcified eggs of the trematode *Schistosoma haematobium* in the kidneys of an Egyptian mummy (3); however, the 1940s and 1950s can be considered the real beginning of paleoparasitology. Several studies were published, providing the first data from Europe (4–8) and South America (9, 10). Then, paleoparasitology continued to develop to suit the interest of scientists throughout the world, and laboratories dedicated to the study of ancient parasites were created first in the Americas and Europe, and more recently in Asia (11–15).

Since the pioneer studies based on mummified material or coprolites, research on intestinal parasites has quickly developed to current analyses performed on various samples, including sediments from latrines and pits,

[1]University of Bourgogne Franche-Comté, Faculty of Sciences and Techniques, CNRS UMR 6249 Chrono-environment, Besançon, France; [2]Escola Nacional de Saude Publica Sergio Arouca, Fundacao Oswaldo Cruz, Rio de Janeiro, Brazil; [†]deceased.
Paleomicrobiology of Humans
Edited by Michel Drancourt and Didier Raoult
© 2016 American Society for Microbiology, Washington, DC
doi:10.1128/microbiolspec.PoH-0013-2015

skeletons, archaeological layers, textiles, and other artifacts in contact with feces remains (16–19). In addition to microscopy, which is still the major tool used by paleoparasitologists, parasites in ancient material are more and more being analyzed with other techniques, such as molecular biology and immunology, which provide access to a larger range of parasites in humans and animals. Evolution in data analysis and, in some cases, the use of statistical tools have allowed the creation of interesting hypotheses and the discovery of answers to technical and epidemiological questions (20–23).

PARASITES, PARASITISM, AND ANCIENT PARASITE EVIDENCE

Although parasitism can develop between almost all living organisms, current research in paleoparasitology is essentially focused on the study of intestinal parasites of humans and animals. Among the parasites developing in the digestive tract, one can distinguish between microparasites and macroparasites (24). In the first group are viruses, bacteria, and protozoa that multiply in the host. The second group includes intestinal worms, or helminths.

All the techniques used to diagnose parasite infections nowadays can also be used to detect parasites in ancient materials (25, 26). Diagnosis in paleoparasitology has evolved from light microscopy diagnosis; light microscopy is now combined with molecular biology and immunology techniques applied to recover evidence of parasite infections in ancient remains (27, 28). Immunology and molecular biology have improved the scope of paleoparasitology in archaeological remains and other kinds of ancient material never examined before (29, 30).

Evidence of past intestinal parasites is based on the recovery of preserved residues in archaeological or paleontological samples of a diversified nature. These residues can be divided into three main groups: macro-

remains; dissemination forms, such as eggs or cysts; and biomolecules, such as antigens or ancient DNA.

Macroremains

In the case of intestinal parasites, macroremains are residues corresponding to the preserved body parts of helminths (adults or larvae). Because of the nature and fragility of the worm body, which is composed only of soft tissues, its preservation requires very favorable conditions, such as a humid, frozen, or dry environment, as well as fossilization. In these conditions, the observation of helminth macroremains is extremely rare. Two cases were reported by Ferreira et al. in which nematode larvae were recovered in coprolites from Brazil (31) and Italy (32). Bouchet and Paicheler (33) identified adult worm remains of *Schistosoma* spp. during a study of latrine samples in Montbeliard (France) by using scanning electronic microscopy. The most sensational record of probable helminth macroremains is the recovery of a fossilized elongated segment of a tapeworm mature proglottid in a shark coprolite dated to 270 million years ago (34). The identified segment contained 93 ovoid eggs and currently represents the oldest occurrence of helminth macroremains and eggs, but more generally currently the most ancient trace of parasitism on earth.

Dissemination Forms of Helminths and Protozoa

This category includes elements such as the eggs of helminths and the cysts of parasite protozoa. The detection of these elements actually represents the most important part of paleoparasitological research activity.

Helminth eggs are round to ovoid elements with an average size between 30 and 160 μm in length, and between 15 and 100 μm in width (35). They consist of a strong multilayer eggshell, partially composed of chitin, which provides good resistance to environmental changes and taphonomic processes. Parasite eggs can be recognized by their morphology (shape, presence of an operculum, ornamentation), and by length and width patterns that allow characterization of the taxon. However, this characterization is often limited to the genus level, with the exception of some very special cases in which the biological origin of the material is known or the morphometrical characteristics of the egg are so precise that a species level can be assigned.

The cysts of parasite protozoans may vary from round to ovoid and have a size ranging from 2 to 90 µm (36). They are much more fragile than helminth eggs and require extreme conditions of preservation, such as a permanently frozen or dry environment. The recovery of ancient cysts is very rare because of this fragility (37, 38), and the study of ancient protozoa requires other techniques, as is explained below.

Biomolecules

Among the parasite biomolecules that have been detected in archaeological samples, the best known are immunological molecules (antigens in particular) and nucleic acids (ancient DNA). Other biomolecules, such as enzymes, have also been detected, but in fewer cases. Antigens are more or less specific macromolecules that are recognized by antibodies. Parasites present a range of antigens, some of which are specific; these can be detected by using adapted immunological tools.

During the last 10 years, immunology has been used to detect protozoan paleoantigens and has made it possible to enlarge the range of parasite infections studied in paleoparasitology. As noted above, protozoan dissemination forms are fragile, and immunology provides an effective method for working on these parasites. Moreover, the specificity of some antigenic markers allows direct identification of the parasites themselves, as well as of their host, leading to a characterization of

the biological origin of the samples. Immunological techniques have been recently introduced, and until now, few analyses have been carried out with commercial or noncommercial kits. However, these techniques are promising for the study of intestinal protozoans. With intestinal parasites, Fouant et al. (39) were the first to try using enzyme-linked immunosorbent assay (ELISA) to confirm the presence of *Entamoeba histolytica* antigens in pre-Columbian mummy samples. Unfortunately, all test results were negative. Some years later, Faulkner et al. (40) obtained the first positive results by using indirect immunofluorescence on a coprolite from the Big Bone Cave in Tennessee (United States) that was radiocarbon dated to 2177 ± 145 BP. Since then, other studies using immunological assays (ELISA, immunofluorescence, immunochromatography) have been done and have successfully recovered intestinal protozoa in archaeological samples (27, 30, 38, 41–46).

Ancient nucleic acids are probably the most promising markers of ancient parasites. The first to be detected were in roundworm eggs (*Ascaris* spp.) preserved in medieval sediments from the site of Namur, Belgium (47, 48). The use of molecular biology has been developed since this period and has resulted in several publications (28, 49–54).

Ancient DNA offers some advantages over classic methods of parasite identification. It provides direct, specific identification, with precise information on the host–parasite relationship, ancient populations, and their environment. Moreover, it allows access to the ancient parasite genome and can provide information on parasite evolution. Dittmar et al. (55) and Dittmar (29) reviewed ancient DNA analyses in paleoparasitology. Because ancient DNA is generally physically and chemically damaged, caution and strict adherence to a sequence of well-established authentication procedures are recommended as the norm (56). Specifically, every experiment should be replicated independently in different laboratories. As Dittmar (29) pointed out, paleoparasitology cannot survive without its root in traditional parasitology. Paleoparasitology is nothing more than an offshoot of parasitology (2).

SAMPLE DIVERSITY AND SAMPLING STRATEGY

Parasite research can be applied to diverse materials. However, the sampling strategy has to be adapted depending on the nature of the sample (16). When one is looking for intestinal worms or protozoa, coprolites are the choice material because they potentially contain parasite dissemination forms. However, because of the leaching phenomenon, parasite forms can be transferred vertically from the coprolite to the sediment. In cases of coprolite sampling, it is important to collect sediment located just under the specimen. Both specimens are considered as a single sample (19, 57, 58).

Body remains such as skeletons are also valuable materials. Studies of skeletons offer the advantage that the biological origin of the material is known (human or animal), which can improve diagnostic ability. In the case of intestinal parasites, sampling is done in the region of the abdominal cavity just near and under the ribs, the vertebrae, and the pelvis. In terms of potential, mummified corpses of humans and animals are probably the most interesting for paleoparasitology because they provide access to soft tissues such as skin, muscles, hair, and even desiccated blood, and they offer the possibility of looking for all kinds of parasites. However, this theoretical approach is often limited by the rarity of such material and the necessity to preserve these precious objects. Imaging techniques have recently been used on mummies to better determine the location of the structures, such as coprolites, to be studied (59).

Paleoparasitological research can also provide information on the function of some shaft features. In this case, the sampling strategy varies according to the type of accumulation (23, 60). In the case of a single layer, sediment samples can be taken from the top, middle, and bottom. In the case of multiple layers, samples should be taken from each layer.

Samples taken from sediment cores allow another type of approach to be adopted. Unlike other types of material, which give information on a short timescale, sediment core samples offer direct access to parasite evolution in one place over time. For example, variations in biodiversity can be observed and put in relation to environmental or cultural changes (61).

Finally, paleoparasitological analyses can be performed in many other contexts. Archaeological layers, funerary urns, embalming jars, shrouds and other mortuary textiles, and potentially all materials or areas in contact with organic remains (fecal matter, putrefaction fluids, etc.) can be studied for intestinal parasites (18, 22).

EXTRACTION METHODS

The use of modern coprological techniques of flotation or concentration is more or less efficient for the recovery of ancient parasite eggs and depends on the nature of the sample, the soil, and taphonomical conditions (17). Parasite extraction follows the same process whatever the nature of the sample (coprolites, sediment samples, textiles, etc.). Some adaptations can be made depending on the type of residues studied. The extraction of helminth eggs requires three successive steps: rehydration, homogenization, and filtration/sieving. The rehydration step (requiring 3 to 10 days on average) derives from a technique with 0.5% trisodium phosphate solution used by Van Cleave and Ross (62). This rehydration technique was used for the first time and introduced into paleoparasitological studies by Callen and Cameron to recover parasite eggs from Peruvian coprolites (63). The second step, homogenization, can be done in different ways: mixing with a simple spoon, crushing the sample in a mortar, using an ultrasonic device during a short time, or combining these different processes. Finally, the samples are filtered in sterile tissue compresses or with a sieving column that has meshes of adapted size (30, 64). In these conditions, microscopic observations can be difficult because of the presence of other organic or mineral elements, sometimes in very high concentrations. Reinhard et al. (17) subjected archaeological samples to palynological treatments and concluded that pollen extraction techniques were not ideal to study helminth eggs from latrine samples. In the same way, Dufour and Le Bailly tested different combination of acids and bases during the rehydration step to try to diminish nonparasitic (organic and mineral) elements and observed a decrease in parasite biodiversity (65). Finally, Anastasiou and Mitchell (66) used distilled water to prepare samples from the Middle East and proposed to simplify the rehydration using only water was compared to the rehydration method using TSP and glycerol. And they decided to adopt the water only method based on the comparisons they made.

Quantification of the residues is a key question for paleoparasitologists that is often debated and tested because it can provide a basis for data comparisons and for epidemiological discussions (17). For example, Herrmann (67, 68) and Jones (23, 69) tested the quantification derived from modern parasitological techniques. A quantitative approach to egg counting was performed by Reinhard and Bryant (70), who aimed to compare parasite infections in different prehistorical groups. This technique was combined with spontaneous sedimentation technique, with the addition of exotic spores (*Lycopodium*) to count eggs (71, 72). Quantification of microfossils is done by adding a known number of exotic *Lycopodium* spores to a known volume or weight of sediment (73).

Difficulties in evaluating parasite loads in eggs found in fecal samples are due to parasite biology and to the fact that the number of eggs eliminated by helminths in a host's feces may vary daily, so that the number of worms in the intestines of the host is difficult to calculate precisely. However, parasite egg counts in the coprolites found in different archaeological sites can be compared as a way to estimate environmental or sanitary conditions (74, 75). In a recent study, Racz et al. applied Reinhard's quantification method to study coprolites directly taken from skeletons recovered in a medieval site in Nivelles, Belgium. They demonstrated a high number of worms (whipworm and roundworm) by calculating the concentration of parasite eggs (76). In these particular conditions of conservation and sampling, quantification can be applied. Difficulties are more significant when the sampling is performed in cesspits in which fecal matter of various origins is probably mixed with soil. For a review of the techniques and methods applied in paleoparasitology, with comments, see Araújo et al. (77).

Antigen research can be done on prepared samples following the standard method previously described. However, the use of additional solution for sample preservation and storage (formalin, for example) can be problematic. Le Bailly and Bouchet (30, 38) proposed rehydrating samples intended for immunological tests with only ultrapure water at 2 to 5°C to avoid the development of fungi or algae. Water rehydration should not interfere with the antibody–antigen link reactions.

Concerning molecular biology techniques, Hofreiter et al., Debruyne and Poinar, and Prufer et al. (78–80) recommended that analysis be started as soon as samples are collected from archaeological sites or selected from collections. This is very important because the less the samples are manipulated, the less contamination will interfere with results. Exogenous DNA is always a problem when one is attempting to recover ancient DNA (81, 82). Kemp and Smith (83, 84) proposed that samples be washed repeatedly in a sodium hypochlorite (NaClO) aqueous solution, then exposed to ultraviolet radiation before molecular biology techniques are used. The laboratory for ancient DNA analyses must be isolated from others, and the same sample must be tested in different places at different times. Molecular targets must be fragments no bigger than 500 bp. Negative controls are required in all steps, but not positive controls. Ancient host DNA must be recovered and authenticated (56).

CONTRIBUTIONS OF PAST INTESTINAL PARASITE STUDIES

Applications in Archaeology

Because intestinal parasites are organisms that may cause disease, the first contribution of paleoparasitology is to acquire information that is relevant to paleopathology. The presence of a parasite in humans and animals may induce negative effects, resulting in varied symptoms, even death, and may be associated with diseases found in skeletons or mummies. Because only a variable part of a population develops symptoms due to the presence of pathogens, the recovery of parasites indicates only the possibility that a disease is linked to these parasites. Many case studies report the presence of parasites in humans or animals and add to the knowledge of disease in ancient populations. Scarce articles also report evidence of intestinal parasite diseases common to humans and animals. This is the case in the study of Dittmar and Teegen (85), who reported the recovery of *Fasciola* spp. eggs dated to 4500 BP in humans and cattle from the German site of Karsdorf. The presence of parasite infection in ancient populations may also shed light on paleoepidemiological aspects of modern diseases. These data supplement those obtained by biological anthropology studies.

The presence of parasites is the result of different ways of life, habits, or behaviors.

Consequently, paleoparasitology can provide information concerning ancient populations, cultures, and environments. Funerary practices, cultural changes, organic refuse management, and hygiene are potential subjects for such studies. In the case of intestinal parasites, because the route of contamination is mainly oral, some conclusions regarding aspects of ancient diet can be reached (74). As an example, from 2001 to 2007, the study of many Neolithic lakeside settlements dated between 3900 and 2900 BC in the northern Alpine region revealed the presence of eggs of several parasites, including whipworms, tapeworms, giant kidney worms, and liver flukes (61) (Fig. 1). Parasite biodiversity showed interesting variations during the thousand-year period under study. In particular, around 3400 BC, an increase in parasites linked to the consumption of fish (biodiversity and frequency) was observed that could be attributed to cultural change, but also to the climate change that occurred during the same period. This colder climate, known as Piora II (Rotmoos II), seems to be at the origin of diet adaptations and forced the local population to adopt growing, gathering, hunting, and fishing activities, the latter probably responsible for the parasite peak (86).

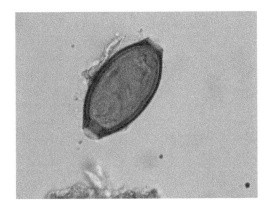

FIGURE 1 **Egg of the whipworm, *Trichuris* spp., with preserved polar plugs, recovered in the archaeological site of Torwiesen II, Germany (55 × 29 μm). (Photo: M. Le Bailly.)**

Paleoparasitological analyses are often carried out on materials and archaeological structures of unknown origin. In the case of intestinal parasites, the recovery of markers will attest to the presence of intestinal remains or fecal matter in a structure. When parasite biodiversity is sufficient, by association or by the consideration of parasites with a specific host spectrum, it is possible to hypothesize about the biological origin of these samples and/or the function of some archaeological structures (60).

Applications in Parasitology and Parasite History

In addition to direct archaeological applications, paleoparasitology aims to retrace parasite history and improve knowledge of the evolution of the host–parasite relationship. By compiling data acquired during different studies, it is possible to obtain a global overview of parasite chrono-geographical evolution. Emergences, disappearances, resurgences, and migrations are some of the observable phenomena often linked to human and/or animal movements (87). Finally, these compilations can offer insights into the future development of parasitic diseases and anticipate their prophylactic defense.

Recently, syntheses were published dealing with intestinal parasites specific to humans. Hookworm (ancylostomid), whipworm (*Trichuris trichiura*), and roundworm (*Ascaris lumbricoides*) infections are known to have originated in humans a long time ago. During human prehistorical migrations, they accompanied their hosts wherever climate condition allowed them to maintain their life cycles. These worms are called geohelminths because part of their life cycle takes place in the soil, where they develop from eggs passed with feces to infective stages at specific temperature (from 22°C to 24°C) and humidity conditions. These intestinal parasites have been found in archaeological sites in different parts of the Old World and New World (11). Because of the cold climate conditions during prehistorical migrations when humans crossed the Bering Land Bridge from Siberia to the Americas, infections with these parasites were lost, and people reached the North American continent free of them. To explain how the three species infected prehistorical populations in North and South America, alternative routes were proposed, such as trans-Pacific routes (88–90).

In the case of *A. lumbricoides*, paleoparasitological studies shed light on an ancient discussion about the origin of this parasite in humans. The debate concerning the origins of the pig roundworm, *Ascaris suum*, and the human roundworm was intense— whether one was transmitted to the other or vice versa. After reviewing the paleoparasitological record and applying molecular biology techniques, Leles et al. (91) concluded that *A. lumbricoides* and *A. suum* are a single species that can infect both humans and pigs, with minor differences in haplotypes (Fig. 2).

Le Bailly and Bouchet (92) synthesized data on the human-specific amoeba responsible for dysentery, *Entamoeba histolytica*. Occurrences of the parasite obtained by immunological analyses were compiled to begin retracing the history of the amoeba. This compilation showed that *E. histolytica* had been present in western Europe since

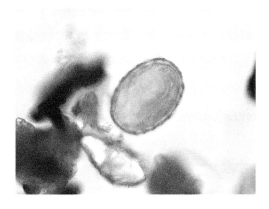

FIGURE 2 **Egg of the roundworm, *Ascaris* spp., recovered in a medieval archaeological site in Laon, France (70 × 45 μm). (Photo: M. Le Bailly.)**

at least the Neolithic and had probably diffused cross Europe, following human movements. As an example, a human pathogenic amoeba from the 12th century was recovered in Israel (45) and could have been imported by Europeans during the Crusades. In the New World, the parasite was detected in the Caribbean and dated to the pre-Columbian period, proving that the parasite was present in the Americas before the 16th century European conquest (93).

A smaller number of compilations dealing with animal parasites have been published. Le Bailly and Bouchet (94) compiled global data on the liver lancet fluke, *Dicrocoelium lanceolatum*, a parasite of herbivores that also can infect humans, although rarely. This compilation showed that the parasite had been present in western Europe since the Paleolithic, then was highly prevalent in France, Switzerland, Germany, and Austria from the Neolithic to the Modern period. The parasite was first detected in North America from the 17th century, a period that corresponds to the arrival of Europeans in Canada.

Similarly, Dufour et al. (95) published a synthesis on the horse pinworm, *Oxyuris equi*, in archaeology and showed that the parasite had been present in central Europe from c. 2500 BP. It seems to have diffused into western Europe by way of the horse migration that accompanied humans, in particular during the Roman period. It finally appeared in Americas relatively recently as a consequence of re-introduction of the horse (Fig. 3).

CONCLUSION

After a century of development and data acquisition, paleoparasitology has become a real bio-archaeological research field, bringing genuine information to archaeology and developing its own problematic on the reconstruction of parasite history and host–parasite evolution. Technological progress

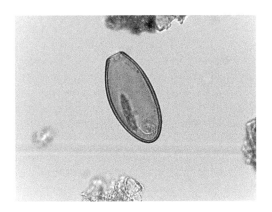

FIGURE 3 **Egg of the horse pinworm,** *Oxyuris equi,* **recovered in the archaeological site of Berel, Kazakhstan (81 × 40 μm). (Photo: B. Dufour.)**

holds promise for the future development of this research field, particularly concerning the detection and identification of parasites—two major points that scientists interested in the evolution of parasites need to explore further. In this regard, development of the use of molecular biology to detect ancient parasites appears to be crucial, and molecular biology would be a good complementary approach for microscopic analysis.

Turning their attention to less-documented regions is another challenge for paleoparasitologists in the future. Some parts of the world, such as the Americas and western European countries, have been well studied for decades. Other regions, such as Asia, northeastern Africa (in particular Egypt and Sudan), and the Middle East, are more recently studied and have been a source of data regularly for the last 10 years (15, 96–98). However, for other areas in the world, information concerning ancient parasites is almost entirely lacking. This is the case, for example, for western and southern Africa, and for eastern and central Europe. Collecting information from these regions is important for a proper understanding of the global history of ancient parasites and could provide comprehensive data to better explain the current distribution of gastrointestinal parasites in the world.

CITATION

Le Bailly M, Araújo A. 2016. Past intestinal parasites. Microbiol Spectrum 4(4):PoH-0013-2015.

REFERENCES

1. **Dittmar K, Araújo A, Reinhard K.** 2012. The study of parasites through time: archaeoparasitology and paleoparasitology, p 170–190. *In* Grauer AL (ed), *A Companion to Paleopathology*, 1st ed. Wiley Blackwell, Hoboken, NJ.
2. **Araújo A, Ferreira LF.** 2000. Paleoparasitology and the antiquity of human host-parasite relationships. *Mem Inst Oswaldo Cruz* **95:**89–93.
3. **Ruffer MA.** 1910. Note on the presence of "*Bilharzia haematobia*" in Egyptian mummies of the twentieth dynasty. *Br Med J* **16:**65.
4. **Szidat L.** 1944. Uber die Erhaltungsfähigkeit von Helmintheneiern in Vor- und Frühgeschichtlichen Moorleichen. *Z Parasitol* **13:**265–274.
5. **Grzywinski L.** 1955. Parasite eggs in faeces originating from the 11th, 12th and 13th centuries. *Pol Tow Parazytol* **4:**97.
6. **Taylor EL.** 1955. Parasitic helminths in mediaeval remains. *Vet Rec* **67:**216–219.
7. **Hoeppli R.** 1956. The knowledge of parasites and parasitic infections from ancient times to the 17th century. *Exp Parasitol* **5:**398–419.
8. **Helbaek H.** 1958. The last meal of Grauballe Man: an analysis of food remains in the stomach. *Kuml:*83–116.
9. **Pizzi T, Schenone H.** 1954. Hallazgo de huevos de *Trichuris trichiura* en contenido intestinal de un cuerpo arqueologico incaico. *Bol Chil Parasitol* **9:**73–75.
10. **Callen EO, Cameron TWM.** 1955. The diet and parasites of prehistoric Huaca Prieta Indians as determined by dried coprolites. *Proc R Soc Canada* **5:**51–52.
11. **Goncalves MLC, Araújo A, Ferreira LF.** 2003. Human intestinal parasites in the past: new findings and a review. *Mem Inst Oswaldo Cruz* **98:**103–118.
12. **Araújo A, Ferreira LF, Fugassa M, Leles D, Sianto L, Mendonça de Souza SM, Dutra J, Iniguez A, Reinhard K.** New World paleoparasitology, p 165–202. *In* Mitchell P (ed), *Sanitation, Latrines and Intestinal Parasites in Past Populations*, Ashgate edition, Surrey, United Kingdom.
13. **Bouchet F, Harter S, Le Bailly M.** 2003. The state of the art of paleoparasitological research in the Old World. *Mem Inst Oswaldo Cruz* **98:**95–101.
14. **Anastasiou A.** 2015. Parasites in European populations from prehistory to the Industrial Revolution, p 203–218. *In* Mitchell P (ed), *Sanitation, Latrines and Intestinal Parasites in Past Populations*, Ashgate edition, Surrey, United Kingdom.
15. **Seo M, Shin DH.** 2015. Parasitism, cesspits and sanitation in the east Asian countries prior to modernisation, p 149,164. *In* Mitchell P (ed), *Sanitation, Latrines and Intestinal Parasites in Past Populations*, Ashgate edition, Surrey, United Kingdom.
16. **Jones AKG.** 1982. Human parasite remains: prospects for a quantitative approach, p 66–70. *In* Hall AR, Kenward HK (ed), *Environmental Archaeology in the Urban Context*. Council for British Archaeology, London, UK.
17. **Reinhard KJ, Confalonieri UE, Herrmann B, Ferreira LF, Araújo AJG.** 1986. Recovery of parasite eggs from coprolites and latrines. *Homo* **37:**217–239.
18. **Harter-Lailheugue S, Bouchet F.** 2006. Palaeoparasitological study of atypical elements of the Low and High Nile Valley. *Bull Soc Pathol Exot* **99:**53–57.
19. **Le Bailly M, Harter S, Bouchet F.** 2003. La paléoparasitologie, à l'interface de l'archéologie et de la biologie. *Archeopages* **11:**12–17.
20. **Confalonieri UEC, Ferreira LF, Araújo AJG, Filho BR.** 1988. The use of statistical test for the identification of helminth eggs in coprolites. *Paleopathol Newsl* **62:**7–8.
21. **Fugassa MH, Araújo A, Guichon RA.** 2006. Quantitative paleoparasitology applied to archaeological sediments. *Mem Inst Oswaldo Cruz* **101:**29–33.
22. **Harter S.** 2003. *Implication de la paléoparasitologie dans l'étude des populations anciennes de la vallée du Nil et du Proche-Orient : études de cas.* Unpublished thesis, Reims Champagne-Ardenne, Reims.
23. **Jones AKG.** 1985. Trichurid ova in archaeological deposits: their value as indicators of ancient faeces, p 105–119. *In* Fieller NJR, Gilbertson DD, Ralph NGA (ed), *Paleobiolgical Investigations: Research Design, Methods and Data Analysis*. British Archaeological Research International Series, Heslington, UK.
24. **Roberts LS, Janovy J Jr.** 2009. *Gerald D. Schmidt & Larry S. Roberts' Foundations of Parasitology*, 8th ed. McGraw-Hill, New-York, NY.
25. **Araújo A, Reinhard K, Bastos OM, Costa LC, Pirmez C, Iniguez A, Vicente AC, Morel CM, Ferreira LF.** 1998. Paleoparasitology: perspectives with new techniques. *Rev Inst Med Trop Sao Paulo* **40:**371–376.

26. **Bouchet F, Guidon N, Dittmar K, Harter S, Ferreira LF, Chaves SM, Reinhard K, Araújo A.** 2003. Parasite remains in archaeological sites. *Mem Inst Oswaldo Cruz* **98**:47–52.

27. **Le Bailly M, Goncalves MLC, Harter-Lailheugue S, Prodeo F, Araújo A, Bouchet F.** 2008. New finding of *Giardia intestinalis* (Eukaryote, Metamonad) in Old World archaeological site using immunofluorescence and enzyme-linked immunosorbent assays. *Mem Inst Oswaldo Cruz* **103**:298–300.

28. **Leles D, Araújo A, Ferreira LF, Vicente ACP, Iniguez AM.** 2008. Molecular paleoparasitological diagnosis of *Ascaris* sp. from coprolites: new scenery of ascariasis in pre-Columbian South America times. *Mem Inst Oswaldo Cruz* **103**:106–108.

29. **Dittmar K.** 2009. Old parasites for a new world: the future of paleoparasitological research. A review. *J Parasitol* **95**:365–371.

30. **Le Bailly M.** 2005. *Evolution de la relation hôte/parasite dans les systèmes lacustres nord alpins au Néolithique (3900-2900 BC), et nouvelles données dans la détection de paléoantigènes de Protozoa.* Thesis, Reims Champagne-Ardenne, Reims.

31. **Ferreira LF, Araújo AJG, Confalonieri UEC.** 1980. The findings of eggs and larvae of parasitic helminths in archaeological material from Unai, Minas Gerais, Brazil. *Trans R Soc Trop Med Hyg* **74**:798–800.

32. **Ferreira LF, Araújo A, Duarte AN.** 1993. Nematode larvae in fossilized animal coprolites from Lower and Middle Pleistocene Sites, Central Italy. *J Parasitol* **79**:440–442.

33. **Bouchet F, Paicheler JC.** 1995. Paleoparasitology: presumption of a case with bilharzia of the 15th century at Montbeliard (Doubs, France). *C R Acad Sci III* **318**:811–814.

34. **Dentzien-Dias PC, Poinar GJ, de Figueiredo AEQ, Pacheco ACL, Horn BLD, Schultz CL.** 2013. Tapeworm eggs in a 270 million-year-old shark coprolite. *PLoS One* **8**:e55007.

35. **Ash LR, Orihel TC.** 2007. *Atlas of Human Parasitology*, 5th ed. ASCP Press, Singapore.

36. **Melhorn H.** 2001. *Encyclopedic Reference of Parasitology*, 2nd ed. Springer Verlag, Berlin, Germany.

37. **Le Bailly M, Leuzinger U, Schlichtherle H, Bouchet F.** 2005. *Diphyllobothrium*: Neolithic parasite? *J Parasitol* **91**:957–959.

38. **Le Bailly M, Bouchet F.** 2006. Paléoparasitologie et immunologie: l'exemple d'*Entamoeba histolytica*. *Archeosciences* **30**:129–135.

39. **Fouant MM, Allison M, Gerszten E, Focacci G.** 1982. Parasitos intestinales entre los indigenas Precolombinos. *Rev Chungara* **9**:285–299.

40. **Faulkner CT, Patton S, Strawbridge-Johnson S.** 1989. Prehistoric parasitism in Tennessee: evidence from the analysis of dessicated fecal material collected from Big Bone Cave, Van Buren County, Tennessee. *J Parasitol* **75**:461–463.

41. **Allison MJ, Bergman T, Gerszten E.** 1999. Further studies on fecal parasites in antiquity. *Am J Clin Pathol* **112**:605–609.

42. **Goncalves ML, Araújo A, Duarte R, da Silva JP, Reinhard K, Bouchet F, Ferreira LF.** 2002. Detection of *Giardia duodenalis* antigen in coprolites using a commercially available enzyme-linked immunosorbent assay. *Trans R Soc Trop Med Hyg* **96**:640–643.

43. **Goncalves ML, da Silva VL, de Andrade CM, Reinhard K, da Rocha GC, Le Bailly M, Bouchet F, Ferreira LF, Araújo A.** 2004. Amoebiasis distribution in the past: first steps using an immunoassay technique. *Trans R Soc Trop Med Hyg* **98**:88–91.

44. **Gonçalves MLC, Schnell C, Sianto L, Bouchet F, Le Bailly M, Reinhard K, Ferreira LF, Araújo A.** 2005. Protozoan infection in archaeological material. *J Biol Res* **80**:146–148.

45. **Mitchell PD, Stern E, Tepper Y.** 2008. Dysentery in the crusader kingdom of Jerusalem: an ELISA analysis of two medieval latrines in the city of Acre (Israel). *J Archaeol Sci* **35**:1849–1853.

46. **Ortega YR, Bonavia D.** 2003. *Cryptosporidium*, *Giardia*, and *Cyclospora* in ancient Peruvians. *J Parasitol* **89**:635–636.

47. **Loreille O, Bouchet F.** 2003. Evolution of ascariasis in humans and pigs: a multidisciplinary approach. *Mem Inst Oswaldo Cruz* **98**:39–46.

48. **Loreille O, Roumat E, Verneau O, Bouchet F, Hänni C.** 2001. Ancient DNA from *Ascaris*: extraction amplification and sequences from eggs collected in coprolites. *Int J Parasitol* **31**:1101–1106.

49. **Iniguez AM, Vicente ACP, Araújo A, Ferreira LF, Reinhard KJ.** 2002. *Enterobius vermicularis*: specific detection by amplification of an internal region of 5S ribosomal RNA intergenic spacer and trans-splicing leader RNA analysis. *Exp Parasitol* **102**:218–222.

50. **Iniguez AM, Reinhard KJ, Araújo A, Ferreira LF, Vicente ACP.** 2003. *Enterobius vermicularis*: ancient DNA from North and South American human coprolites. *Mem Inst Oswaldo Cruz* **98**:67–69.

51. **Iniguez AM, Reinhard K, Goncalves MLC, Ferreira LF, Araújo A, Vicente ACP.** 2006. SL1 RNA gene recovery from *Enterobius vermicularis* ancient DNA in pre-Columbian human coprolites. *Int J Parasitol* **36**:1419–1425.

52. **Jaeger LH, Iniguez AM.** 2014. Molecular paleoparasitological hybridization approach as effective tool for diagnosing human intestinal parasites from scarce archaeological remains. *Plos One* **9:**e105910. doi:10.1371/journal.pone.0105910.

53. **Leles D, Cascardo P, Freire Ados S, Maldonado A Jr, Sianto L, Araújo A.** 2014. Insights about echinostomiasis by paleomolecular diagnosis. *Parasitol Int* **63:**646–649.

54. **Shapiro B, Hofreiter M.** 2012. *Ancient DNA: Methods and Protocols, Methods in Molecular Biology.* Springer, Berlin, Germany.

55. **Dittmar K, de Souza SM, Araújo A.** 2006. Challenges of phylogenetic analyses of aDNA sequences. *Mem Inst Oswaldo Cruz* **101:**9–13.

56. **Willerslev E, Cooper A.** 2005. Ancient DNA. *Proc Biol Sci* **272:**3–16.

57. **Dommelier-Espejo S.** 2001. *Contribution à l'étude paléoparasitologique des sites néolithiques en environnement lacustre dans les domaines jurassien et péri-alpin.* Unpublished Ph.D. dissertation, University of Reims Champagne-Ardenne, Reims, France.

58. **Le Bailly M, Bouchet F.** 2006. La paléoparasitologie: les parasites comme marqueurs de la vie des populations anciennes. *Studii de Preistorie* **3:**225–232.

59. **Mendonça de Souza S.** 2008. Diagnóstico não Invasivo em Múmias Milenares, p 77–99. *In* Heron W, Jr, Lopes J (ed), *Paleontologia. Arqueologia. Fetologia.* Tecnologias 3D. Revinter, Rio de Janeiro, Brazil.

60. **Herrmann B.** 1988. *Parasite Remains from Mediaeval Latrine Deposits: an Epidemiologic and Ecologic Approach.* Note et Monographie Technique No. 24, p 135–142. CNRS, Paris.

61. **Le Bailly M.** 2011. *Les parasites dans les lacs nord alpins au Néolithique (3900-2900 BC) et nouvelles données dans la détection des paléoantigènes de Protozoa.* Editions Universitaires Européennes, Sarrebrück, Germany.

62. **Van Cleave HJ, Ross JA.** 1947. A method of reclaiming dried zoological specimens. *Science* **105:**318.

63. **Callen EO, Cameron TWM.** 1960. A prehistoric diet revealed in coprolites. *New Scientist* **8:**35–40.

64. **Bouchet F, West D, Lefèvre C, Corbett D.** 2001. Identification of parasitoses in a child burial from Adak Island (Central Aleutian Islands, Alaska). *C R Acad Sci* **324:**123–127.

65. **Dufour B, Le Bailly M.** 2013. Testing new parasite egg extraction methods in paleoparasitology and an attempt at quantification. *Int J Paleopathol* **3:**199–203.

66. **Anastasiou E, Mitchell PM.** 2013. Simplifying the process of extracting intestinal parasite eggs from archaeological sediment samples: A comparative study of the efficacy of widely-used disaggregation techniques. *Int J Paleopathol* **3:**204–207.

67. **Herrmann B.** 1986. Parasitologische Untersuchungen eines Spätmittelalterlich-Frühneuzeitlichen Kloakeninhaltes aus der Fronerei auf dem Schrangen in Lübeck. *Lübecker Schriften zur Archäologie und Kulturgeschichte* **12:**167–172.

68. **Herrmann B.** 1987. Parasitologische Untersuchung mittelalterlicher Kloaken, p 160–169. *In* Herrmann B (ed), *Mensch und Umwelt im Mittelalter.* Deutsche Verlags-Anstalt, Stuttgart, Germany.

69. **Jones AKG.** 1986. Parasitological Investigations on Lindow Man, p 136–139. *In* Stead IM, Bourke JB, Brothwell D (ed), *Lindow Man, the Body in the Bog.* Book Club Associates, London, UK.

70. **Reinhard K, Bryant VM, Jr.** 1992. Coprolites analysis: a biological perspective on archaeology, p 245–288. *In* Schiffer MB (ed), *Advances in Archaeological Method and Theory,* vol 4. Academic Press, New York, NY.

71. **Sianto L, Chame M, Silva CSP, Goncalves MLC, Reinhard K, Fugassa M, Araújo A.** 2009. Animal helminths in human archaeological remains: a review of zoonoses in the past. *Rev Instit Med Trop São Paulo* **51:**119–130.

72. **Camacho M, Pessanha T, Leles D, Dutra J, Silva R, Mendonça de Souza SM, Araujo A.** 2013. Lutz' spontaneous sedimentation technique and the paleoparasitological analysis of sambaqui (shellmounds) sediments. *Mem Inst Oswaldo Cruz* **108:**155–159.

73. **Reinhard K, Damon T, Edwards SK, Meier DK.** 2006. Pollen concentration analysis of ancestral Pueblo dietary variation. *Palaeogeogr Palaeoclimatol Palaeoecol* **237:**92–109.

74. **Reinhard K, Ferreira LF, Bouchet F, Sianto L, Dutra J, Iniguez AM, Leles D, Le Bailly M, Fugassa M, Pucu E, Araújo A.** 2013. Food, parasites, and epidemiological transitions: A broad perspective. *Int J Paleopathol* **86:**150–157.

75. **Reinhard K, Pucu E.** 2014. Comparative parasitological perspectives on epidemiologic transitions: the Americas and Europe, p 321–336. *In* Zuckerman MK (ed), *Modern Environments and Human Health.* Wiley Blackwell, John Wiley and Sons, Hoboken, New Jersey.

76. **Rácz SE, Pucu de Araújo E, Jensen E, Mostek C, Morrow JJ, Van Hove ML, Bianucci R, Willems D, Heller F, Araújo A, Reinhard KJ.** 2015. Parasitology in an archaeological context: analysis of medieval burials in Nivelles, Belgium. *J Archaeol Sci* **53:**304–315.

77. **Araújo A, Reinhard K, Ferreira LF.** 2015. Palaeoparasitology – human parasites in ancient material. *Adv Parasitol*, in press.
78. **Hofreiter M, Serre D, Poinar HN, Kuch M, Pääbo S.** 2001. Ancient DNA. *Nat Rev Genet* **2:**353–359.
79. **Debruyne R, Poinar HN.** 2009. Time dependency of molecular rates in ancient DNA data sets, a sampling artifact? *Syst Biol* **58:**348–360.
80. **Prufer K, Stenzel U, Hofreiter M, Paabo S, Kelso J, Green RE.** 2010. Computational challenges in the analysis of ancient DNA. *Genome Biol* **11:**47.
81. **Drancourt M, Raoult D.** 2005. Palaeomicrobiology. Current issues and perspectives. *Nat Rev Microbiol* **3:**23–35.
82. **Hebsgaard MB, Phillips MJ, Willerslev E.** 2005. Geologically ancient DNA: fact or artefact? *Trends Microbiol* **13:**212–220.
83. **Kemp BM, Smith DG.** 2005. Use of bleach to eliminate contaminating DNA from the surface of bones and teeth. *Forensic Sci Int* **154:**53–61.
84. **Kemp BM, Smith DG.** 2010. Ancient DNA methodology: thoughts from Brian M. Kemp and David Glenn Smith on "Mitochondrial DNA of protohistoric remains of an Arikara population from South Dakota." *Hum Biol* **82:**227–238.
85. **Dittmar K, Teegen WR.** 2003. The presence of *Fasciola hepatica* (liver fluke) in humans and cattle from a 4,500 year old archaeological site in the Saale-Unstrut Valley, Germany. *Mem Inst Oswaldo Cruz* **98:**141–143.
86. **Le Bailly M, Leuzinger U, Schlichtherle H, Bouchet F.** 2007. "Economic crash" during the Neolithic at the Pfyn-Horgen transition (3400 BC): contribution of the paleoparasitology. *Anthropozoologica* **42:**175–185.
87. **Araújo A, Reinhard K, Ferreira LF, Gardner S.** 2008. Parasites as probes for prehistoric migrations? *Trends Parasitol* **24:**112–115.
88. **Araújo A, Ferreira LF, Confalonieri U, Chame M.** 1988. Hookworms and the peopling of America. *Cadernos de Saude Publica* **2:**226–233.
89. **Ferreira LF, Araujo A.** 1996. On hookworms in the Americas and trans-Pacific contact. *Parasitol Today* **12:**454–454.
90. **Montenegro A, Araújo A, Eby M, Ferreira LF, Hetherington R, Weaver AJ.** 2006. Parasites, paleoclimate, and the peopling of the Americas. *Current Anthropol* **47:**193–200.
91. **Leles D, Gardner SL, Reinhard K, Iniguez AM, Araújo A.** 2012. Are *Ascaris lumbricoides* and *Ascaris suum* a single species? *Parasit Vectors* **5:**42.
92. **Le Bailly M, Bouchet F.** 2015. A first attempt to retrace the history of dysentery caused by *Entamoeba histolytica*, p 219–228. *In* Mitchell P (ed), *Sanitation, Latrines and Intestinal Parasites in Past Populations*, Ashgate ed. Surrey.
93. **Le Bailly M, Romon T, Kacki S.** 2014. New evidence of *Entamoeba histolytica* infections in pre-Columbian and colonial cemeteries in the Caribbean. *J Parasitol* **100:**684–686.
94. **Le Bailly M, Bouchet F.** 2010. Ancient dicrocoeliosis: occurences, distribution and migration. *Acta Trop* **115:**175–180.
95. **Dufour B, Hugot J-P, Lepetz S, Le Bailly M.** 2015. The horse pinworm (*Oxyuris equi*) in archaeology during the Holocene: review of past records and new data. *Infect Genet Evol* **33:**77–83.
96. **Nezamabadi M, Harter-Lailheugue S, Le Bailly M.** 2011. Paleoparasitology in the Middle-East: state of the research and potential. *Tüba-Ar* **14:**205–213.
97. **Nezamabadi M, Aali A, Stöllner T, Mashkour M, Le Bailly M.** 2013. Paleoparasitological analysis of samples from the Chehrabad salt mine (northwestern Iran). *Int J Paleopathol* **3:**229–233.
98. **Mowlavi G, Mokhtarian K, Makki MS, Mobedi I, Masoumian M, Naseri R, Hoseini G, Nekouei P, Mas-Coma S.** 2015. *Dicrocoelium dendriticum* found in a Bronze Age cemetery in western Iran in the pre-Persepolis period: the oldest Asian palaeofinding in the present human infection hottest spot region. *Parasitol Int* **64:**251–255.

Paleopathology and Paleomicrobiology of Malaria

15

ANDREAS NERLICH[1]

Malaria, one of the deadliest diseases of humankind, remains a major global health problem in the 21st century (1, 2). In 2014, 198 million persons were infected, with more than 0.5 million deaths from malaria globally. Malaria is recognized as the second leading cause of death from infectious diseases in Africa, after HIV/AIDS, and is the fifth most frequent cause of death from infectious diseases worldwide, after respiratory infections, HIV/AIDS, diarrheal diseases, and tuberculosis (2).

Malaria is caused by endoparasites of the genus *Plasmodium* (3). The infection is transmitted to humans through the bites of female anopheles flies. Therefore, the distribution of the disease is mostly restricted to the habitat of the vector, which preferentially covers the tropics and subtropics but which was spread far beyond this area in historical times. Today, we distinguish five subtypes of *Plasmodium* that are pathogenic to humans and lead to different clinical features. *P. falciparum* causes severe malaria with high fever (malaria tropica). Infection by *P. knowlesi* also mostly has a severe and often lethal course, with a febrile cycle of approximately 24 hours, whereas infection with *P. malariae*, *P. vivax*, or *P. ovale* has a less severe clinical course with the typical manifestations of malaria tertiana and quartana (i.e., fever episodes of 2 or 3 days' duration). The latter three types of malaria are mostly associated with the undulating fever that is so typical of the infection. However, it should

[1]Institute of Pathology, Academic Clinic Munich-Bogenhausen, Munich, Germany.
Paleomicrobiology of Humans
Edited by Michel Drancourt and Didier Raoult
© 2016 American Society for Microbiology, Washington, DC
doi:10.1128/microbiolspec.PoH-0006-2015

be remembered that the most severe, tropical form (malaria tropica) does not have this clinical feature.

The bite of an infected anopheles fly usually transports the parasite into the human body, where the plasmodia invade hepatocytes and subsequently red blood cells. The latter are then used as a "reproduction shelter" so that the parasites can multiply without being directly accessible to the host's immune system. After the completion of parasite reproduction and thereby multiplication, infected erythrocytes break up and numerous new parasites spread into the bloodstream, where they rapidly invade more red blood cells. The rupture of the erythrocytes liberates cytokines that lead to a rapidly increasing fever in the host; thus, a synchronous replication cycle of some of the plasmodia (e.g., *P. malariae*, *P. vivax*, and *P. ovale*) results in the typical undulating fever of 2 to 3 days' duration.

Taking the typical clinical picture into account, we can date early literary evidence of malaria infection back to the early Greek period, when Hippocrates described the typical undulating fever (4) that is highly suggestive of malaria. However, ancient Egyptian papyri, such as the Papyrus Ebers, do not provide us with typical descriptions of recurrent fever; rather, they describe numerous conditions with high and/or irregular fevers that may have resulted from the tropical form of malaria infection but also from many other infections (5). Images of insects (e.g., in ancient Egyptian temple walls) may suggest the presence of anopheles flies, although the correct iconographic attribution of these images remains uncertain. Interestingly, both bees and anopheles flies have only three pairs of legs; the hieroglyph, however, depicts five legs, so that a secure "diagnosis" seems impossible (Fig. 1).

IDENTIFICATION OF MALARIA INFECTION IN PALEOPATHOLOGY

Malaria infection does not lead to specific or even diagnostically characteristic traces

FIGURE 1 **Temple wall representation of an insect (Temple of Queen Hatchepsut, Deir-el-Bahari, Thebes-West, Egypt, c. 1300 BC). Most Egyptological references translate this hieroglyph as "bee"; however, there is also some potential resemblance to anopheles flies.**

in osseous paleopathological material, such as are seen in other chronic infections like tuberculosis and syphilis. However, chronic recurrent infection and destruction of the red blood cells results in chronic anemia, which may manifest in the skeleton as typical hyperostotic lesions of the cranial vault (porotic hyperostosis) or the orbita (cribra orbitalia) (Fig. 2) (6). Previous anthropological and paleopathological studies, such as the one by Gowland and Western (7), applied a combination of geospatial and biomorphological data to suggest that malaria was indeed prevalent in marshy areas of Anglo-Saxon England (AD 410 to 1050), obviously caused by *P. vivax* infections. However, because both types of bone lesions are also induced by other anemic conditions (e.g., recurrent blood loss in chronic intestinal infections, dietary lack of iron, inherited anemias), neither porotic hyperostosis nor cribra orbitalia necessarily indicates chronic malaria. Furthermore, the acute forms of the infection (especially malaria tropica) often have a very rapid clinical course, so that the loss of erythrocytes does not manifest in the skeleton.

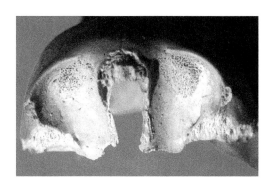

FIGURE 2 **Macropathological example of severe chronic anemia evidenced by orbital pitting (cribra orbitalia). Similar morphological changes may occur in chronic anemia caused by malaria. However, cribra orbitalia and other porotic hyperostoses of the skull are also seen in chronic deficiency conditions, including anemia with other causes.**

In some instances, macroscopic osseous lesions are suggestive of inherited hemolytic anemias, such as sickle cell anemia and thalassemia, with characteristic spicular new bone formation of the cranium (6). These inborn diseases of erythropoiesis are much more frequently seen in areas with a history of endemic malaria and have been regarded as an adaptive response to the malarial environment (6). Therefore, the identification of such paleopathological conditions may be an indirect indication of endemic malaria.

Rarely, the paleopathological material consists of mummified corpses or fragments thereof, which may contain well-preserved soft tissues that can be used for histomorphological investigation. Although such studies have occasionally identified erythrocytes that survived dehydration and/or artificial embalming until now (8), there is no report on the identification of malarial parasites within erythrocytes.

Modern paleopathology therefore has adopted molecular techniques – comparable with those of modern clinical infectiology – in order to unambiguously identify infection by plasmodial parasites. Accordingly, two general types of approaches have been developed that either prove the presence of specific parasitic proteins or identify the presence of specific plasmodial ancient DNA (aDNA) segments (9). Both approaches have been used successfully, and the results provide initial insight into the presence and epidemiology of malaria infection, mainly in ancient Egypt.

Immunological Identification of Malaria

In 1994, Miller and co-workers were the first to use customized dipstick tests for the plasmodia-specific "histidine rich protein-2 antigen" (10) to identify malaria infections in soft-tissue extracts from ancient Egyptian mummies. This approach was later pursued on a larger scale by Rabino Massa et al. (11) in mummies from the Torino Museum collection. In this series, more than 40% of the analyzed cases reacted positively in the immunological tests. Approximately 92% of the positive cases revealed signs of chronic anemia (see above). Later studies in an Early Dynastic mummy (12) and Renaissance period European skeletal remains (from the famous de' Medici family) confirmed positive results (13, 14). Accordingly, several members of the de' Medici family had malaria, as evidenced by a positive immunological reaction. Most remarkably, the immunological test of the latter studies was conducted with a modern test set that obviously showed far less false-positive cross-reaction, although occasional test results were obviously falsely negative (15). Taken together, these studies made it clear that the careful application of this technique reduces the risk for false-positive or nonspecifically reacting cases, and that the technique may provide an important screening method to identify cases with probable infection by plasmodia. Finally, the immunological tests do not distinguish between any substrains of plasmodia; for example, the *CareStart* Malaria HRP-2/pLDH Combo Test identifies the malaria histidine rich type-2 protein (HRP-2) of *P. falciparum*, but also the lactate dehydrogenase (pLDH) of *P. vivax*, *P. ovale*, and *P. malariae* (15).

Ancient DNA Analysis of Malaria

Since the very first molecular study of malaria in paleopathology, conducted by Taylor et al. in 1997 on a case from approximately 60 years ago (16), several study groups have confirmed the usefulness of this approach to identify malaria infections in paleopathology. Likewise, in 2001, Sallares and Gomzi (17) identified among 40 skeletons from a place in Lugnano, Umbria (central Italy) one with a specific *P. falciparum* sequence of 18S rDNA with 98% homology to present-day sequences. The affected individual was a 2- to 3-year-old child from the late Roman period (5th century AD).

In parallel, our group analyzed skeletal material from an ossuary in southern Germany dating to 1400 to 1800 AD; one case among 20 individuals with evidence of chronic anemia (see above) showed a positive result. The infected individual had a *Plasmodium*-specific aDNA sequence of 18S rDNA with 98% homology to present plasmodial sequences (18).

In a further methodical approach, our group successfully identified two individuals infected by *P. falciparum* in a series of 91 ancient Egyptian skeletonized mummies from a broad time range between 3500 and 500 BC. Both positive cases came from the Tombs of the Nobles of the large necropolis of Thebes-West, Upper Egypt, and dated back to approximately 1550 to 1000 BC (19). One was an adult male and the other an adult female, both of high social rank. During the extensive analytical procedure in this study, the *P. falciparum* chloroquine-resistance transporter (pfcrt) gene proved to produce the most reliable results, with 99% homology to the present-day plasmodial gene segment. Data were confirmed in two independent laboratories.

In 2010, Hawass et al. (20) reported their paleopathological and molecular findings in the mummy of Tutankhamun and several other presumed ancient Egyptian royal family members. In addition to numerous other observations, the authors included a test for the *Plasmodium*-specific sequences of the AMA1 gene (*P. falciparum*) and were able to identify among 13 persons five who had been infected by *P. falciparum*. These individuals again dated back to between 1550 and 1324 BC, and the tests were repeated in two independent laboratories for confirmation.

Finally, in 2014, Lalremruata et al. (21) performed a further molecular study on 16 mummy heads from Lower Egypt (Abusir el-Meleq, Fayum, 1064 BC to 300 AD) and applied two *P. falciparum*–specific gene segments (AMA1, MSP1). In this small series, six individuals tested positive for malaria tropica, which was confirmed by sequence analysis. Interestingly, the study also tested for tuberculosis (*Mycobacterium tuberculosis* complex aDNA) and showed a high co-incidence with four co-infected individuals in the series.

Besides the direct identification of plasmodial aDNA, indirect evidence of malaria infection may come from the analysis of human genetic mutations that are associated with resistance to malaria (22). In their extensive review of the history of malaria in southern Europe, Sallares et al. (22) describe increased frequencies of thalassemia, sickle cell anemia, and glucose-6-phosphate dehydrogenase (G6PD) deficiency in certain Mediterranean populations and concluded that the increased frequencies of these conditions were the result of recurrent malaria epidemics with high pressure on genetic selection mechanisms.

Problems and Limitations in the Detection of Plasmodia in Paleopathology

Because the paleopathological inspection of skeletal remains allows the detection of only "insecure" (i.e., noncharacteristic) traces of chronic plasmodial infection (those of chronic anemia), it seems mandatory that modern paleopathologists use immunological and/or molecular biological techniques in order to verify malaria infections in history before speculating on the presence of malaria (and eventually on its frequency) in historical populations.

For routine purposes, modern analytical techniques have been adapted to freshly taken blood samples, but they have also proved useful in dried blood samples. Nevertheless, the postmortem decomposition of proteins and/or DNA may destroy molecular targets and lead to "false-negative" results; however, cross-reaction with nonspecific targets may also lead to "false-positive" results. In this regard, all analytical procedures must comply with well-established guidelines for working with ancient protein and DNA (23); moreover, scientists should always try to combine techniques and to use information from other sources (e.g., the presence of vector remains in historical samples) to enhance the plausibility of their observations.

Furthermore, the data available thus far indicate that only plasmodia of the subspecies *P. falciparum* have been identified – none of the other strains. In this regard, it is of note that the immunological approach does not necessarily distinguish among the various strains (see above), whereas the species-specific PCR amplicons suggested by Taylor et al. (16) reveal multiple nonspecific products. We await more complex analyses, such as whole-genome analysis, to identify the occurrence of the various *Plasmodium* substrains – information that it is hoped will shed light on the evolutionary pathways of human pathogenic plasmodia.

CONCLUSION

Recent immunological and molecular biological techniques have unambiguously confirmed the presence of malaria over a period of at least about 3,000 years. Although the number of positively tested cases is still low – at present not more than 15 cases have been identified by aDNA analysis – the small numbers of samples tested overall suggest fairly high rates of infection by the parasites.

Assuming that ongoing studies will provide us with more data on this issue, we can expect that malaria has had considerable impact on historical populations, at least in the Near East and Europe. It is hoped that both the evolutionary aspects of the host–pathogen interaction and the spread of the disease will be uncovered in the near future.

CITATION

Nerlich A. 2016. Paleopathology and paleomicrobiology of malaria. Microbiol Spectrum 4(3):PoH-0006-2015.

REFERENCES

1. **Murray CJL, Rosenfeld LC, Lim SS, Andrews KG, Foreman KJ, Haring D, Fullman N, Naghavi M, Lozano R, Lopez AD.** 2012. Global malaria mortality between 1980 and 2010: a systematic analysis. *Lancet* **379:**413–431.
2. **World Health Organization.** 2014. *World Malaria Report 2014.* World Health Organization, Geneva, Switzerland. http://www.who.int/malaria/publications/world_malaria_report_2014/wmr-2014-no-profiles.pdf?ua=1.
3. **White NJ, Pukrittayakamee S, Hien TT, Faiz MA, Mokuolu OA, Dondorp AM.** 2014. Malaria. *Lancet* **383:**723–35.
4. **Bogdonoff MD, Crellin JK, Good RA, McGovern JP, Nuland SB, Saffon MH.** 1985. *The Genuine Work of Hippocrates* (Hippocrates, Epidemics 1.6,7,24–26; Aphorisms 3.21,22;4.59,63; On Airs, Waters and Places c. 10). Classics of Medicine Library, Birmingham, AL.
5. **Ebers G.** 1875. Papyros Ebers. Das hermetische Buch über die Arzneimittel der Alten Ägypter. W. Engelmann Verlag, Leipzig, Germany.
6. **Aufderheide AC, Rodriguez-Martin C.** 2005. *The Cambridge Encyclopedia of Human Paleopathology.* Cambridge University Press, Cambridge, UK.
7. **Gowland RL, Western AG.** 2012. Morbidity in the marshes: using spatial epidemiology to investigate skeletal evidence for malaria in Anglo-Saxon England (AD 410-1050). *Am J Phys Anthropol* **147:**301–311.
8. **Stout SD, Teitelbaum SL.** 1976. Histological analysis of undecalcified thin sections of archeological bones. *Am J Phys Anthropol* **44:**263–269.
9. **Anastasiou E, Mitchell PD.** 2013. Palaeopathology and genes: investigating the genetics of infectious diseases in excavated human

skeletal remains and mummies from past populations. *Gene* **528:**33–40.

10. **Miller RL, Ikram S, Armelagos GJ, Walker R, Harer WB, Schiff CJ, Baggett D, Carrigan M, Maret SM.** 1994. Diagnosis of *Plasmodium falciparum* infections in mummies using the rapid manual ParaSight-F test. *Trans R Soc Trop Med Hyg* **88:**31–32.

11. **Rabino Massa E, Cerutti N, Marin D, Savoia A.** 2000. Malaria in ancient Egypt: paleoimmunological investigations in predynastic mummified remains. *Chungara* **32:**7–9.

12. **Bianucci R, Mattutino G, Lallo R, Charlier PH, Jouin-Spriet H, Peluso A, Higham T, Torre C, Rabino Massa E.** 2008. Immunological evidence of *Plasmodium falciparum* infection in a child mummy from the Early Dynastic Period. *J Archaeol Sci* **35:**1880–1885.

13. **Fornaciari G, Giuffra V, Ferroglio E, Gino S, Bianucci R.** 2010. Malaria was "the killer" of Francesco I de' Medici (1531-1587). *Am J Med* **123:**568–569.

14. **Fornaciari G, Giuffra V, Ferroglio E, Gino S, Bianucci R.** 2010. *Plasmodium falciparum* immunodetection in bone remains of members of the Renaissance Medici family (Florence, Italy, sixteenth century). *Trans R Soc Trop Med Hyg* **104:**583–587.

15. **Bianucci R, Tognotti E, Giuffra V, Fornaciari G, Montella A, Milanese M, Floris R, Bandiera P.** 2014. Origins of malaria and leishmaniasis in Sardinia: first results of a paleoimmunological study. *Pathologica* **106:**89.

16. **Taylor GM, Rutland P, Molleson T.** 1997. A sensitive polymerase chain reaction method for the detection of *Plasmodium* species DNA in ancient human remains. *Ancient Biomol* **1:**193–204.

17. **Sallares R, Gomzi S.** 2001. Biomolecular archaeology of malaria. *Ancient Biomol* **3:**195–213.

18. **Zink A, Haas CJ, Herberth K, Nerlich AG.** 2001. PCR amplification of *Plasmodium* DNA in ancient human remains. *Ancient Biomol* **3:**293.

19. **Nerlich AG, Schraut B, Dittrich S, Jelinek TH, Zink A.** 2008. *Plasmodium falciparum* in Ancient Egypt. *Emerg Infect Dis* **14:**1317–1318.

20. **Hawass Z, Elleithy H, Gad YZ, Ismail S, Hasan N, Ahmed A, Amer H, Khairat R, Fathalla D, Ball M, Pusch CM, Gaballah F, Wasef S, Fateen M, Gostner P, Selim A, Zink A.** 2010. Ancestry and pathology in King Tutankhamun's family. *JAMA* **303:**638–647.

21. **Lalremruata A, Ball M, Bianucci R, Welte B, Nerlich AG, Kun JFJ, Pusch CM.** 2013. Molecular identification of falciparum malaria and human tuberculosis co-infections in mummies from the Fayum Depression (Lower Egypt). *PlosOne* **8:**e60307. doi:10.1371/journal.pone.0060307.

22. **Sallares R, Bouwman A, Anderung C.** 2004. The spread of malaria to Southern Europe in antiquity: new approaches to old problems. *Med History* **48:**311–328.

23. **Cooper A, Poinar H.** 2000. Ancient DNA: do it right or not at all. *Science* **289:**1139.

History of Smallpox and Its Spread in Human Populations

16

CATHERINE THÈVES,[1] ERIC CRUBÉZY,[1] and PHILIPPE BIAGINI[2]

Smallpox, the infectious disease caused by species of the variola virus (VARV), is probably one of the most terrible diseases to have affected human populations over the past hundreds of years. Its dissemination was significantly related to global population growth and the movement of people across regions and continents. The geographical origin of the disease remains a matter of debate; hypotheses suggest the Indus Valley or Egypt and the Near East, regions that had high population densities 3,000 to 4,000 years ago (1, 2). The latter hypothesis was recently refined by Babkin and Babkina (3), who suggested that the initial spread of the virus in humans could have occurred in the Horn of Africa (Kingdom of the Queen of Sheba). In this region, active trade expeditions overlapped with the distribution areas of several animal poxvirus hosts (including the naked-soled gerbil) and the introduction of the domesticated camel as a new potential host. The disease spread from these regions to the west and east, with historical reports suggesting epidemics in China and Europe as early as the 1st or 2nd century CE, with a subsequent progressive emergence of the virus in western Africa (4). During the 16th century, smallpox was a significant cause of death in Europe. The smallpox agent was also exported to South America during this time and passed over

[1]AMIS Laboratory, UMR 5288, CNRS / University of Toulouse / University of Strasbourg, Toulouse, France;
[2]Viral Emergence and Co-Evolution Unit, UMR 7268 ADES, Aix-Marseille University / French Blood Agency / CNRS, Marseille, France.

Paleomicrobiology of Humans
Edited by Michel Drancourt and Didier Raoult
© 2016 American Society for Microbiology, Washington, DC
doi:10.1128/microbiolspec.PoH-0004-2014

both American continents (5, 6). As noted by Fenner et al., smallpox was a major global endemic disease by the mid-18th century, with the exception of Australia (4). Only the variolation and vaccination campaigns initiated more than two centuries ago reduced dramatically the spread and impact of the disease in contemporary populations.

HISTORY OF SMALLPOX

From Ancient Times to 1000 CE

According to the hypotheses reported above, it is possible to locate VARV in ancient Egypt. Literature details three mummies dated from the 18th and 20th dynasties (1580 to 1350 BCE and 1200 to 1100 BCE, respectively) whose skin was covered with lesions that resembled those of the smallpox rash (7). Thus, the earliest physical evidence of a VARV infection consists of the scars identified on these mummies, including the well-known mummy of the Egyptian Pharaoh Ramses V, currently preserved in the Cairo Museum, who died in 1157 BCE. Molecular studies aiming to confirm the smallpox diagnosis have not been authorized. In addition, extensive writings from that period describing the devastating effects of the disease and/or specific medical subjects have not been identified. The first description of smallpox in Egypt was reported by Aaron of Alexandria in 622 CE (4, 8). Smallpox could have been brought to the East by Egyptian traders traveling by caravan or across the sea during the 1st millennium BCE.

Descriptions of diseases potentially related to smallpox outbreaks in Ethiopian, Persian, and Syrian populations were retrieved in writings dating from c. 300 to c. 900 CE; however, it is accepted that the clearest description of smallpox was made by the Persian physician Al-Razi, who differentiated the disease from measles in c. 910 CE (4, 8). In regard to India, some authors postulated that the disease was mentioned in the oldest Sanskrit texts, associated with the recognition of a dedicated Hindu

god, *Sitala* (1, 8, 9); however, a precise description of the disease, *masurika*, was reported only around the 7th century CE by the physician Vagbhata (4). The period when smallpox was introduced into China is also under debate; a precise dating (i.e., between 25 CE and 49 CE) was reported by the physician Ko Hung in 340 CE (10), whereas some authors suggested a valid knowledge of the disease as early as the 11th century BCE (8). The subsequent spread of the virus to Japan, between the 6th and 7th centuries CE, was most likely favored by cultural (Buddhism) and trading exchanges with China and Korea. The erection of a great statue of Buddha (*Nara Daibutsu*, 752 CE) was commissioned by the Emperor Shomu at the end of the smallpox epidemic in memory of this period.

The dissemination of the disease in Europe and western Africa was suggested initially in writings dating from the 5th and 6th centuries CE. It is conceivable that the introduction of smallpox in European populations would have been linked mainly to war and invasion events. Thus, the invasion of the Huns around 451 CE would have been contemporaneous with the possible presence of smallpox in French populations, as noted by Saint Nicaise, Bishop of Reims at that time. Despite a dearth of information on smallpox for the remainder of the Middle Ages, it is assumed that the Moorish invasions during the 7th and 8th centuries CE contributed to the spread of the disease in southwestern Europe (4). Whether smallpox was also involved in several well-known historical episodes, including the Plague of Athens (Peloponnesian Wars, 430 BCE) and the Antonine Plague (165 CE), when smallpox could have been carried by Roman soldiers returning to Syria and Italy from Mesopotamia, are interesting hypotheses that remain to be elucidated (4–11).

Spread of Smallpox during the Second Millennium CE

During the 11th and 13th centuries CE, the Crusades contributed vastly to the dissemination of smallpox in Europe. Interestingly, an analysis of historical sources of Constantinople reports two cases that, in all probability, correspond to smallpox (12). The spread of smallpox was persistent during the subsequent centuries, and by the 1500s, the disease was considered endemic in most European countries, from Iceland to Spain. Royal houses were also affected, as exemplified by the illness of Queen Elizabeth I of England in 1562, who remained disfigured (8). The introduction of the disease in the New World, in the early 16th century, was linked to the creation of Spanish and Portuguese colonies. As a consequence of Conquistador expeditions and wars against the Aztec and Inca empires, it has been estimated that smallpox killed approximately 3 million to 4 million inhabitants (6). During the same period, another source of importation of the virus was the contamination of ports in West Africa; thus, smallpox spread throughout Central America via the slave trade (13). Populations in Peru and Brazil were subsequently devastated (c. 1524 and c. 1555, respectively), and by the beginning of the 17th century, smallpox outbreaks had been identified throughout South America (4). The first epidemics in North America occurred in 1617 to 1619 (Massachusetts) as a result of colonization by European settlers; other ports, including Boston and New York, suffered major epidemics during the 17th century.

Eyler noted that smallpox replaced bubonic plague as Europe's most feared disease at that time (14). The disease also became established in Russia (1623) and was disseminated across the country, with massive epidemics in Siberian populations (1630) (4).

The death of Queen Mary II of England (1694) was symbolic of the future devastations that were to be observed during the 18th century. Smallpox affected all levels of society and killed an estimated 400,000 Europeans each year, in a population of approximately 160 million people in 1750 (13, 14). Several other monarchs also succumbed: Emperor Joseph I of Austria, King Louis XV

of France, Tsar Peter II of Russia, King Luis I of Spain, and Queen Ulrika Eleonora of Sweden. The disease appeared in southern Africa in 1713 (Cape Town), and by the end of the 18th century, smallpox was considered a major global endemic disease after its introduction into the Australian continent in 1789.

Following Edward Jenner's discovery in 1796, variolation and, subsequently, more efficient vaccination protocols that were developed during the 19th century contributed to reduce progressively the incidence of the disease. Severe epidemic episodes were, however, still observed in the 1800s (e.g., in European and North American cities), and the disease was also present during the Franco-Prussian War of 1870 to 1871 (8).

Smallpox was still endemic in many countries at the beginning of the 20th century (4). The situation evolved favorably after the Second World War and the subsequent decisive decision of the WHO in 1959 to initiate an eradication program targeting smallpox. The application of the WHO strategic action plan, the Intensified Smallpox Eradication Programme (1967 to 1980), resulted in mass vaccination campaigns and the development of surveillance systems. By 1979, the WHO officially declared smallpox eradicated (1, 14).

PALAEOMICROBIOLOGY AND PALAEOGENETIC STUDIES

Palaeomicrobiology is the study of disease-causing microorganisms in skeletal and mummified remains from archaeological contexts. Although initial microscopy studies were conducted on cross-sections of mummy tissues as early as 1889 (15), the development of genetic studies involving historical or archaeological specimens was relatively recent and mainly dependent on PCR technology (16). More recently, advances in microbiological knowledge permitted the detection of pathogens from the past (17). Fragments of bacterial or viral DNA are investigated in possibly infected tissues, or in the teeth of subjects who died following bacteremia or viremia events, with the specific characteristics of the ancient DNA samples taken into account (18, 19). The technical requirements for palaeomicrobiology studies are the same as those for human palaeogenetic analyses (20). The possibility of contamination in archaeological samples occurs during excavation (soil environment) and/or during experimental analyses in a laboratory dedicated to ancient DNA. Therefore, contamination controls must be conducted to validate the detection of ancient pathogens in investigated samples and in the authentication of past diseases (21, 22).

Studies of Historical Cases

Mummies are human remains, relatively ancient and relatively complete or incomplete. All are highly dehydrated. The preservation of different tissues can be exceptional and can reflect the pathological signs of disease outbreaks that have affected human populations over the centuries. If pathogens can be characterized by using laboratory tests, diseases may be better understood, especially if they can be integrated into a historical context, thus offering new insights into disease epidemiology. Several mummies showing evidence of variola infection have been reported; the combined ages of these mummies spans thousands of years of history, as described below.

Ruffer is considered a pioneer of histopathology (15). He described, with his colleague Ferguson, a skin rash on an Egyptian mummy dated from the 12th dynasty (1200 to 1100 BCE) whose distribution, pattern, and type closely resembled those of smallpox (7). Histological sections showed vesicles with vertical septa, consistent with the appearances observed in this disease. This first diagnosis was contested, however, as the pathological characteristics of smallpox could not be robustly confirmed (23–25).

Early research noted that the most famous mummy, that of the Egyptian Pharaoh

Ramses V (1157 BCE) presented characteristics of a smallpox-like rash (26). Research aiming to demonstrate smallpox virus particles in the vesicles was not completely conclusive. Hopkins (27) was the first to study the distribution of the rash, although not all body parts could be examined. He noted that "[the] rash is quite striking and is remarkably similar to smallpox." Subsequent studies of skin pieces by electron microscopy were inconclusive because no virus particles were observed in the cells (27). Radioimmunoassay analysis of skin vesicles from the shroud of the royal mummy gave negative results, and the immunological reactivity of the viral particles was weak (28). Moreover, electron microscopic observations resulted in the identification of a single, morphologically atypical virus particle (28, 29). Strouhal (30) used the results described above with those from a mummification study to argue that the burial of the pharaoh, and other members of the royal family, took place quickly because of a possible smallpox epidemic. Currently, there are no biological tests to conclude that Ramses V had a smallpox infection.

At the same time, Fornaciari and Marchetti (31) published a description of a widespread eruption on a 400-year-old Italian mummy, again with macroscopic aspects and a smallpox-like regional distribution. A more complete study was performed, including electron microscopic analysis of the lesions, and confirmed virus-like structures with a size and morphology corresponding to those of orthopoxviruses. Viral antigenic activity and DNA molecular hybridization tests were negative as the virus was no longer viable (32).

In 1985, archaeologists led excavations in London church crypts where burials had taken place between 1726 and 1856, a period when smallpox was endemic in Europe (4). The cool, dry conditions in the crypts could have promoted the good preservation of bodies in wooden coffins. It was questioned whether in these conditions a live virus could still be present on the skin of smallpox victims. Indeed, in one coffin, archaeologists

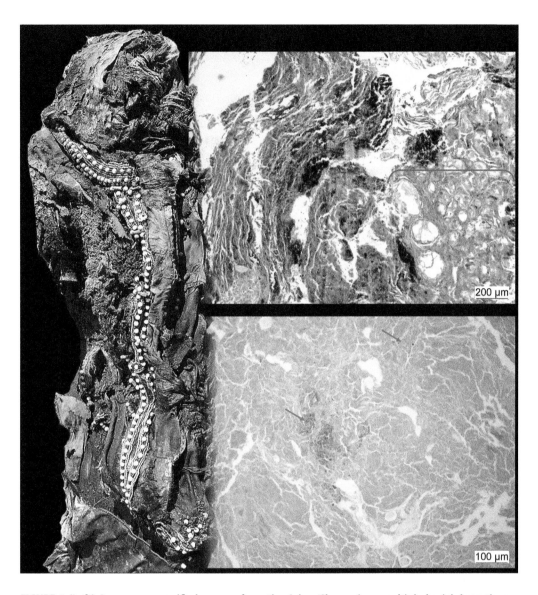

FIGURE 1 **(Left) A young mummified woman from the Arbre Chamanique multiple burial; lung tissues were studied by microscopy and DNA analysis. (Courtesy of Patrice Gérard, CNRS.) (Top right) The lung tissue structure was identified. It exhibited numerous scattered, black-pigmented deposits, potentially the pigment of black lung disease (coal worker's pneumoconiosis). Hematoxylin and eosin coloration, ×50. Courtesy of Catherine Cannet, IML, Strasbourg.) (Bottom right) A significant amount of iron was found in the lung parenchyma and may correspond to important bleeding. Perls coloration, × 100. Courtesy of Catherine Cannet, IML, Strasbourg.)**

found a desiccated piece of skin with small-pox lesions, bound to a skeleton. This skin sample from a 100-year-old body underwent electron microscopic testing, but no viable virus was detected (33). Following this work,

almost 60 bodies with partially preserved skin were analyzed, but no smallpox lesions were identified.

Human remains can be naturally mummi-fied by cold temperatures in areas where

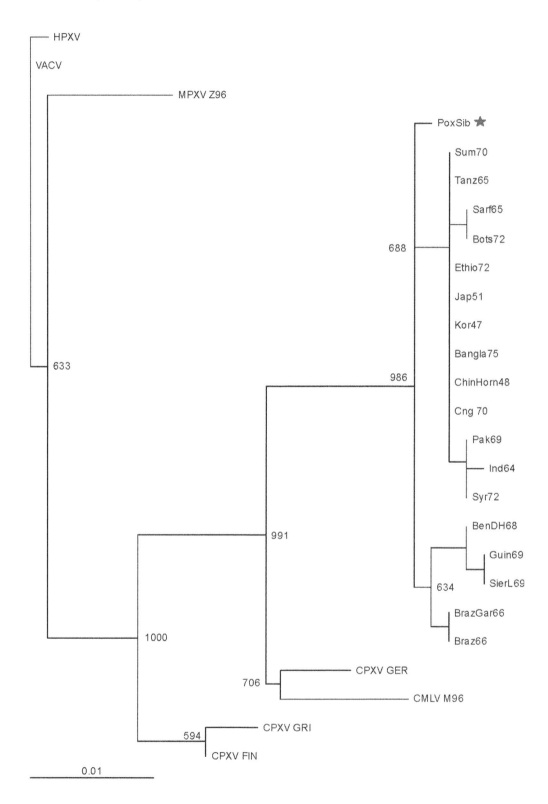

permafrost remains in the ground. In the warm July of 1991, in the village of Yakutia above the Arctic Circle, a temperature change of the permafrost led to the appearance of the mummified bodies of smallpox victims from the 19th century (34). Scientists from the State Research Center of Virology and Biotechnology VECTOR (Novosibirsk, Russia) and their Yakutian colleagues exhumed one mummified child with characteristic smallpox pustules on the skin. After sampling, laboratory analyses did not isolate a live virus; it is possible that the permafrost temperature changes eliminated viruses in the tissues.

Archaeological excavations were undertaken in central Yakutia, and a multiple burial from the 18th century was discovered in 2004 (35). The human genetic analysis of five subjects demonstrated that they belonged to the same family (36). A disease outbreak could be hypothesized because the bodies did not have traces of injuries. Although none of the subjects showed any traces of blisters or pustules on their skin, the hypothesis of a disease outbreak was reinforced by the presence of iron in the lung of the best-preserved subject, suggesting death due to hemorrhage (Fig. 1). A lung sample from this subject was analyzed in a laboratory dedicated to ancient DNA. Three DNA fragments of the ancient human poxvirus were amplified by PCR (37). The absence of pustules, presence of a pulmonary hemorrhage, and detection of the virus indicated that the subject had hemorrhagic smallpox. The first phylogenetic study conducted on these fragments (with a concatenated alignment of 718 bp) showed that this ancient strain was distinct from modern smallpox strains described previously (Fig. 2) (5, 37). As for the earlier report (34), despite the presence of permafrost in Yakutia, in which well-preserved mummified remains are observed after being buried for two and a half centuries, the smallpox DNA was fragmented and degraded. Studies have concluded that Yakutia was struck by several smallpox outbreaks from the 17th to 19th centuries, with significant mortality (38–40).

In 2004, Schoepp and colleagues described the discovery of a child's forearm and hand showing characteristics of a smallpox infection (41). The specimen was dated to be approximately 50 years old. Pathological symptoms were analyzed by microscopy and showed the morphological features of smallpox lesions. Different stages of immature as well as mature virus particles were identified by electron microscopic analysis. Real-time PCR assays (TaqMan technology) were used to detect variola-specific DNA and *Orthopoxvirus* species; the positive amplicon produced in the variola-specific assay was sequenced, and the subsequent phylogenetic analysis confirmed its close relationship with variola major sequences available in databases at that time (41). It is important to note that these studies detected a human smallpox strain in an unknown subject (41) and identified an ancient viral sequence not belonging to human smallpox lineages of the 20th century (37).

In 2011, during construction in Queens, New York City, a well-preserved female mummy dating from the 19th century was

FIGURE 2 Phylogenetic analysis of concatenated sequences of an ancient Siberian smallpox virus (PoxSib) and representative strains identified in humans and animals. Phylogeny suggests that the 300-year-old viral sequence did not belong to the cluster of strains sequenced from the 20th century (1947 to 1975). Strains identified in humans: Guin69, Guinea 1969; SierL69, Sierra Leone 1969; BenDH68, Benin 1968; BrazGar66 and Braz66, Brazil 1966; Syr72, Syria 1972; Pak69, Pakistan 1969; Ind64, India 1964; Sarf65, South Africa 1965; Bots72, Botswana 1972; Ethio72, Ethiopia 1972; Bangla75, Bangladesh 1975; Sum70, Sumatra 1970; ChinHorn48, China 1948; Kor47, Korea 1947; Jap51, Japan 1951; Tanz65, Tanzania 1965; Cng70, Congo 1970. Strains identified in animals: CMLV M96, camelpox virus M96; CPXV FIN, cowpox Finland 2000; CPXV GER, cowpox GER91-3; CPXV GRI, cowpox GRI-90; HPXV, horsepox virus 76; MPXV Z96, monkeypox Zaire-96; VACV, vaccinia virus Copenhagen-derived clone 1990. (Derived from Fig. 1B in reference 37. Reproduced with permission.)

found in an iron coffin (42). The forensics team recorded smallpox-like lesions on the body. Analyses by the Centers for Disease Control and Prevention in Atlanta, Georgia (USA) revealed that the smallpox DNA was overly degraded, with no trace of the virus.

Mummified tissues are not the only ones in which the smallpox virus can be identified. An exceptional case of osteomyelitis secondary to smallpox was found in human remains in a Native American cemetery dating from 1640 to 1650 CE (43). Based on radiological and osteological literature, the author showed that smallpox has an effect on bones; these very specific changes are called osteomyelitis variolosa. The afflicted subject, found in an archaeological context, had severe changes to his bones that could have been the result of contracting smallpox after puberty; surviving the infection led to pathological alterations. Moreover, historical records describe an epidemic in 1634 in Ontario (43); smallpox was introduced along with other diseases after the arrival of settlers (6).

Scab samples from patients in the variolation program of the 18th century were investigated by McCollum and colleagues, although live viruses were not detected (25). Arita (44) demonstrated that specimens collected 9 months before analysis were no longer positive. The author proposed that "variolation material probably becomes inactive" at 1 year after sampling.

GENETICS AND EVOLUTION

Members of the genus *Orthopoxvirus* (family *Poxviridae*) are among the largest and most complex viruses. Approximately 200 genes are distributed across conserved (central) or variable (terminal) regions; their linear DNA genome is double-stranded, ranging from approximately 180 to 230 kbp, with covalently closed ends (45). In contrast to essential genes (involved in virus replication), which are located on the central portion of the genome, flanking terminal regions code for proteins harboring sequence diversity, including those modulating host range and virulence (2, 46, 47). Orthopoxviruses are widespread in vertebrates; they have been identified in human populations and various animal hosts, including camels, cows, mice, and monkeys. The VARV possesses the smallest genome (~186 kbp, ~150 genes). Sequence comparisons of smallpox strains identified in humans have characterized two clades: clade I, with variants of variola major, and clade II, composed of West African variants of variola major and variola minor (alastrim). Variola major and variola minor differ greatly in their mortality rates (~30% vs. ~1%, respectively) and can be distinguished by comparison of their respective sequences (46, 48, 49). Estimates of mutation rates have been based, up to now, on the analysis of complete/partial sequences originating from biological samples dating from five to six decades ago. Accordingly, the mutation accumulation rate of poxviruses is estimated at 0.5×10^{-6} to 7×10^{-6} nucleotide substitutions per site per year (2, 50, 51). The *Orthopoxvirus* phylogeny indicates that three species – VARV, CMLV (camelpox), and TATV (taterapox infecting naked-soled gerbils) – have a common ancestor, presumably a rodent virus (Fig. 3), and that VARV is a relatively young virus (~3000 to ~4000 years old) (Fig. 3), consistent with a high degree of lethality for humans (50, 52, 53). In a recent study, Babkin suggested that the Horn of Africa could have been the overlapping site of the distribution areas of VARV, CMLV, and TATV hosts, leading to emergence of the virus in human populations (3). Considering that smallpox is a specifically human infectious disease, knowledge of human migrations and related historical events is of importance when VARV evolution is analyzed and dated. As detailed above, the hypothesis of a VARV origin in the African continent is the most plausible, although an Indian origin of the virus has also been proposed (2, 3).

Following the initial infectious event, the spread of the virus was closely related to

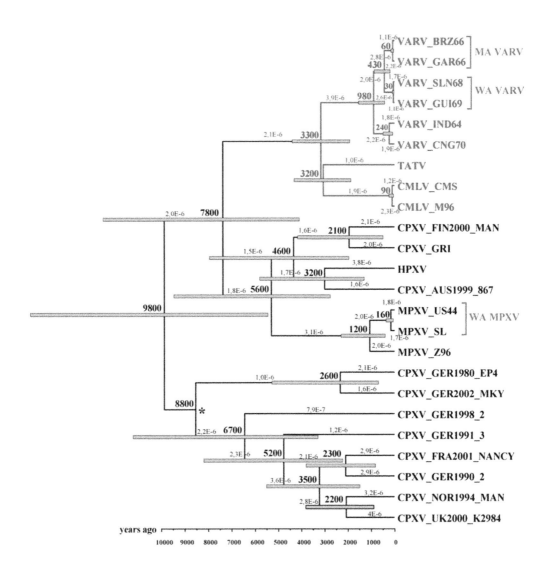

FIGURE 3 Phylogenetic tree built with central conserved regions of diverse orthopoxvirus genomes (maximum credibility tree, Bayesian method) (53). Mutation accumulation rates are shown (substitutions/site/year). Numbers on nodes indicate the time to the most recent common ancestor of the clades (in years). Gray bars: 95% highest probability density intervals; the posterior probabilities of all clades are >90% except the node marked with an asterisk. Strains: VARV, variola virus; TATV, taterapox virus; CMLV, camelpox virus; CPXV, cowpox virus; HPXV, horsepox virus; MPXV, monkeypox virus; MA VARV, variola minor alastrim strains; WA VARV, West African variola virus strains; WA MPXV, West African monkeypox virus strains. (Legend derived from Fig. 2 in reference 53; figure reproduced from Fig. 2 in reference 53 with permission.)

human migrations, trading, and wars, leading to the current description of distinct variants of VARV. Phylogenetic analysis demonstrates that available strains can be grouped into several clusters according to the sampling loca-

tion – that is, Asian, African (other than West African), West African, and South American (5). Interestingly, the latter cluster corresponds to variola minor variants (alastrim) for which a dating has been proposed, from the 1500s to the 1800s, consistent with an exportation of the virus from West Africa to South America during the slave trade (2, 3, 5). Recently, the direct identification of VARV sequences (PoxSib strain) from the frozen tissues of a Siberian mummy dating from the early 1700s made it possible, for the first time, to gain insights about an ancient strain of the virus and about historical epidemic events contemporaneous with the Russian conquest of this region (37, 40). Interestingly, the characterization of short partial sequences suggested that PoxSib could be a direct progenitor of modern viral strains or a member of an ancient lineage that was not the cause of outbreaks in the 20th century (Fig. 2). Such recent findings reinforced the evident interest in palaeomicrobiology studies dedicated to smallpox and to microbes in general.

CONCLUSION

Variola virus is a relatively young virus that emerged approximately 3,000 to 4,000 years ago, following, very probably, its transmission from an animal host and adaptation to a human host. The spread of smallpox, considered among the most devastating of human diseases, was linked to the progressive development of contact between human populations, wars and conquests, and increases in population densities. Smallpox can therefore be considered a disease of civilization. There are numerous historical descriptions of smallpox, but at present only a few written reports dating from the middle to late 1st millennium CE are considered robust. In contrast, the spread of the disease during the 2nd millennium CE, especially from the 16th to the 19th centuries, has been described unequivocally. The recent development of molecular technologies has accelerated our knowledge of the natural history of smallpox; such advances can now be transposed to the analysis of ancient material so that past VARV infections and their correlations with ancient writings can be better understood.

CITATION

Thèves C, Crubézy E, Biagini P. 2016. History of smallpox and its spread in human populations, Microbiol Spectrum 4(4):PoH-0004-2014.

REFERENCES

1. **Geddes AM.** 2006. The history of smallpox. *Clin Dermatol* **24:**152–157.
2. **Shchelkunov SN.** 2009. How long ago did smallpox virus emerge? *Arch Virol* **154:**1865–1871.
3. **Babkin IV, Babkina IN.** 2012. A retrospective study of the orthopoxvirus molecular evolution. *Infect Genet Evol* **12:**1597–1604.
4. **Fenner F, Henderson DA, Arita I, Jezek Z, Ladnyi ID.** 1988. *Smallpox and Its Eradication.* World Health Organization, Geneva, Switzerland.
5. **Li Y, Carroll DS, Gardner SN, Walsh MC, Vitalis EA, Damon IK.** 2007. On the origin of smallpox: correlating variola phylogenics with historical smallpox records. *Proc Natl Acad Sci U S A* **104:**15787–15792.
6. **Mann CC.** 2007. *1491: New Revelations of the Americas Before Columbus.* Vintage Books, New York, NY.
7. **Ruffer MA, Ferguson AR.** 1911. Note on an eruption resembling that of variola in the skin of a mummy of the twentieth dynasty (1200-1100 BC). *J Pathol Bacteriol* **15:**1–3.
8. **Behbehani AM.** 1983. The smallpox story: life and death of an old disease. *Microbiol Rev* **47:**455–509.
9. **Nicholas RW.** 1981. The goddess Sitala and epidemic smallpox in Bengal. *J Asian Stud* **41:**21–45.
10. **Needham J, Lu GD.** 1988. Smallpox in history. *In* Needham J (ed), *Science and Civilization in China,* **vol 6,** part 4. Cambridge University Press, London, UK.
11. **Sabbatini S, Fiorino S.** 2009. The Antonine Plague and the decline of the Roman Empire. The role of the Pathian and Marcomanni Wars between 164 and 182 AD in spreading contagion. *Le Infezioni in Medicina* **4:**261–275.
12. **Lascaratos J, Tsiamis C.** 2002. Two cases of smallpox in Byzantium. *Int J Dermatol* **41:**792–795.
13. **Riedel S.** 2005. Edward Jenner and the history of smallpox and vaccination. *Proceedings (Baylor University Medical Center)* **18:**21–25.
14. **Eyler JM.** 2003. Smallpox in history: the birth, death, and impact of a dread disease. *J Lab Clin Med* **142:**216–220.

15. **Sandison AT.** 1967. Sir Marc Armand Ruffer (1859–1917) pioneer of palaeopathology. *Med Hist* **11:**150–156.

16. **Pääbo S, Higuchi RG, Wilson AC.** 1989. Ancient DNA and the polymerase chain reaction. The emerging field of molecular archaeology. *J Biol Chem* **264:**9709–9712.

17. **Drancourt M, Raoult D.** 2005. Palaeomicrobiology: current issues and perspectives. *Nat Rev Microbiol* **3:**23–35.

18. **Green RE, Krause J, Ptak SE, Briggs AW, Ronan MT, Simons JF, Du L, Egholm M, Rothberg JM, Paunovic M, Pääbo S.** 2006. Analysis of one million base pairs of Neanderthal DNA. *Nature* **444:**330–336.

19. **Thèves C, Senescau A, Vanin S, Keyser C, Ricaut FX, Alekseev AN, Dabernat H, Ludes B, Fabre R, Crubézy E.** 2011. Molecular identification of bacteria by total sequence screening: determining the cause of death in ancient human subjects. *PLoS One* **6:**e21733. doi:10.1371/journal.pone.0021733.

20. **Anastasiou E, Mitchell PD.** 2013. Palaeopathology and genes: investigating the genetics of infectious diseases in excavated human skeletal remains and mummies from past populations. *Gene* **528:**33–40.

21. **Cooper A, Poinar HN.** 2000. Ancient DNA: do it right or not at all. *Science* **289:**1139.

22. **Willerslev E, Cooper A.** 2005. Ancient DNA. *Proc Biol Sci* **272:**3–16.

23. **Sandison A.** 1967. Diseases of the skin, p 449–456. *In* Brothwell D, Sandison A (ed), *Diseases in Antiquity.* Charles C Thomas, Springfield, IL.

24. **Sandison AT.** 1972. Evidence of infective disease. *J Hum Evol* **1:**213–224.

25. **McCollum AM, Li Y, Wilkins K, Karem KL, Davidson WB, Paddock CD, Reynolds MG, Damon IK.** 2014. Poxvirus viability and signatures in historical relics. *Emerg Infect Dis* **20:**177–184.

26. **Eliott Smith G.** 1912. *The Royal Mummies.* Gerald Duckworth & Co. Ltd., London, UK.

27. **Hopkins DR.** 1980. Ramses V: earliest known victim? *World Health* 220.

28. **Lewin P.** 1984. "Mummy" riddles unravelled. *Bull Elect Micro Soc Canada* 3–8.

29. **Lewin PK.** 1988. Technological innovations and discoveries in the investigation of ancient preserved man. *Zagreb Paleopathology Symp* 90–91.

30. **Strouhal E.** 1996. Traces of a smallpox epidemic in the family of Ramesses V of the Egyptian 20th dynasty. *Anthropologie* **34:**315–319.

31. **Fornaciari G, Marchetti A.** 1986. Intact smallpox virus particles in an Italian mummy of sixteenth century. *Lancet* **2:**625.

32. **Marennikova S, Shelukhina E, Zhukova O, Yanova N, Loparev V.** 1990. Smallpox diagnosed 400 years later: results of skin lesions examination of 16th century Italian mummy. *J Hyg Epidemiol Immunol* **34:**227–231.

33. **Baxter PJ, Brazier AM, Young SE.** 1988. Is smallpox a hazard in church crypts? *Br J Ind Med* **45:**359–360.

34. **Stone R.** 2002. Is live smallpox lurking in the Artic? *Science* **295:**2002.

35. **Crubézy E, Alexeev A.** 2007. Chamane: Kyys, jeune fille des glaces. Errance, Paris, France.

36. **Keyser C, Hollard C, Gonzalez A, Fausser JL, Rivals E, Alexeev AN, Ribéron A, Crubézy E, Ludes B.** 2015. The ancient Yakuts: a population genetic enigma. *Philosophical Transactions of the Royal Society B* **370.** doi:10.1098/rstb.2013.0385.

37. **Biagini P, Thèves C, Balaresque P, Géraut A, Cannet C, Keyser C, Nikolaeva D, Gérard P, Duchesne S, Orlando L, Willerslev E, Alekseev AN, de Micco P, Ludes B, Crubézy E.** 2012. Variola virus in a 300-year-old Siberian mummy. *N Engl J Med* **367:**2056–2058.

38. **Tokarev SA.** 1945. Obshestvennyui stroi jakutov XVII-XVIII, (Social Order of Iakuts XVII-XVIII) [In Russian], p 395–396. Jakut. gos. izd-vo, Iakoutsk.

39. **Zinner EP.** 1968. Sibir v izvaestiaukh zapadnoevropeiskikh pouteshestvennikov i outhsenykh XVIII v, (Siberia by travellers of the eighteen century) [in Russian], p 54. *The book of Eastern siberia.* Vost.-Sib. kn. izd-vo. Irkutsk.

40. **Thèves C, Biagini P, Crubézy E.** 2014. The rediscovery of smallpox. *Clin Microbiol Infect* **20:**210–218.

41. **Schoepp RJ, Morin MD, Martinez MJ, Kulesh DA, Hensley L, Geisbert TW, Brady DR, Jahrling PB.** 2004. Detection and identification of Variola virus in fixed human tissue after prolonged archival storage. *Lab Invest* **84:**41–48.

42. **Reardon S.** 2014. Infectious diseases: smallpox watch. *Nature* **509:**22–24.

43. **Jackes MK.** 1983. Osteological evidence for smallpox: a possible case from seventeenth century Ontario. *Am J Phys Anthropol* **60:**75–81.

44. **Arita I.** 1980. Can we stop smallpox vaccination? *World Health* 27–29. http://www.who.int/iris/handle/10665/202496.

45. **Lefkowitz EJ, Wang C, Upton C.** 2006. Poxviruses: past, present and future. *Virus Res* **117:**105–118.

46. **Esposito JJ, Sammons SA, Frace AM, Osborne JD, Olsen-Rasmussen M, Zhang M, Govil D, Damon IK, Kline R, Laker M, Li Y, Smith GL, Meyer H, Leduc JW, Wohlhueter RM.** 2006. Genome sequence diversity and clues to the evolution of Variola (smallpox) virus. *Science* **313:**807–812.

47. **Shchelkunov SN.** 1995. Functional organization of variola major and vaccinia virus genomes. *Virus Genes* **10:**53–71.

48. **Babkin IV, Nepomniashchikh TS, Maksiutov RA, Gutorov VV, Babkina IN, Shchelkunov SN.** 2008. Comparative analysis of variable regions in the genomes of variola virus [in Russian]. *Mol Biol (Mosk)* **42:**612–624.

49. **Hughes AL, Irausquin S, Friedman R.** 2010. The evolutionary biology of poxviruses. *Infect Genet Evol* **10:**50–59.

50. **Babkin IV, Shelkunov SN.** 2008. [Molecular evolution of poxviruses [in Russian]. *Genetika* **44:**1029–1044.

51. **Babkin IV, Babkina IN.** 2011. Molecular dating in the evolution of vertebrate poxviruses. *Intervirology* **54:**253–260.

52. **Shchelkunov SN.** 2011. Emergence and re-emergence of smallpox: the need for development of a new generation smallpox vaccine. *Vaccine* **29:**D49–D53.

53. **Babkin IV, Babkina IN.** 2015. The origin of the variola virus. *Viruses* **7:**1100–1112.

17

Cholera

DONATELLA LIPPI,[1] EDUARDO GOTUZZO,[2] and SAVERIO CAINI[3]

CHOLERA: A HISTORY OF PANDEMICS

Cholera is an acute and often fatal disease of the gastrointestinal tract. In its typical epidemic form, it presents with profuse watery diarrhea and often leads to dehydration and eventually the death of an untreated patient within a few hours (1). The causative agent of cholera is a bacterium known as *Vibrio cholerae*, two serogroups of which (O1, to which the El Tor biotype belongs, and O139) have epidemic potential and are also responsible for endemic cholera. *V. cholerae* has two known reservoirs: humans (who can also be asymptomatic carriers) and the aquatic environment (both freshwater, such as that of the Ganges River delta in India, and the sea). Humans mostly are infected through contaminated water used for drinking or preparing foods, and (when symptomatic) they keep shedding bacteria with feces during 1 to 2 weeks (1). In the aquatic reservoir, *V. cholerae* can persist indefinitely and undergo genetic modification, which makes the eradication of cholera unlikely to be achieved, if not impossible (1).

Cholera epidemics typically last 5 to 10 years; the epidemic that began in Peru in 1991, for instance, ended in less than 10 years, as did most cholera

[1]Department of Experimental and Clinical Medicine, University of Florence, Florence, Italy; [2]Institute of Tropical Medicine, Peruvian University Cayetano Heredia, Lima, Peru; [3]Cancer Risk Factors and Lifestyle Epidemiology, Cancer Research and Prevention Institute (ISPO), Florence, Italy.
Paleomicrobiology of Humans
Edited by Michel Drancourt and Didier Raoult
© 2016 American Society for Microbiology, Washington, DC
doi:10.1128/microbiolspec.PoH-0012-2015

epidemics in Latin America during the 19th century. Cholera epidemics are usually limited by a lack of favorable conditions for survival of the vibrios. It is only where specific ecological conditions in aquatic environments are met that the vibrios can survive indefinitely and sustain an endemic presence of the disease. These conditions include, among others, moderate salinity, warm temperatures, and the presence of copepods, plankton, and molluscs, all of which are typical features of estuaries and coastal swamps like Bengal Bay and the Ganges River (2).

It is now widely admitted that cholera originated in Asia, particularly in India (3). European explorers of many nationalities (Portuguese at the beginning, then Dutch, French, and British) gave several descriptions of epidemics likely ascribable to cholera between 1503 and 1817. The typical symptoms of cholera (vomiting and diarrhea) are common to many other diseases with a similar mode of transmission, which made it difficult to discriminate it from other diseases that ravaged Asia, especially India. Despite these uncertainties, it seems reasonable to believe that the disease was present in this region since at least the 16th century. The year 1817, however, marks the debut of a new era in the history of cholera (4). Cholera had hitherto been a disease mostly localized to a well-defined region of the world, despite some outbreaks that had already been described outside Asia, especially Europe, since the 16th century (4). In 1817, a particularly intense cholera epidemic burst upon India, probably favored by abnormal weather conditions. After wreaking havoc in the country, the disease spread into several other regions of the world, despite the vain attempts of governments to contain its fury. The outbreak spread initially to Nepal and then headed east, touching southeastern Asia, Indonesia, and Borneo, and finally China and Japan in 1821 and 1822. In the same years, cholera also invaded the Middle East, reaching Syria and the southwestern part of present-day Russia in 1823. The particularly severe winter of 1823 to 1824 probably contributed to ending the transmission of the disease in 1824, which thus receded before reaching Europe.

The epidemic of 1817 to 1824, originally referred to as "Asiatic cholera," is nowadays known as the first cholera pandemic. After this first appearance, the disease struck again in several other major pandemics during the 19th and 20th centuries, frequently following trade routes or the movements of the army troops and eventually reaching all regions of the world (5). The second (1829 to 1851) through the sixth (1899 to 1923) pandemics began in India, subsequently spreading to Russia, western Europe, the Middle East, the Americas, and northern Africa. The latest cholera pandemic originated in Indonesia in the 1960s and is still ongoing, although much diminished. Caused by the El Tor biotype of *V. cholerae*, it has reached virtually all regions of the world, thanks to modern means of transport and the frequent mass migrations caused by war and famine.

THE 1854 CHOLERA EPIDEMICS IN LONDON AND THE BIRTH OF MODERN EPIDEMIOLOGY

Cholera holds a special place among the infectious diseases that have accompanied the history of humankind; during the cholera outbreak that took place in Soho, London, in 1854, the English physician John Snow (1813 to 1858) for the first time used modern epidemiological methods to trace the source of the outbreak, describe its time course, understand the causes, and take effective measures to stop its spread (6, 7). John Snow's investigations during the 1854 cholera outbreak in London (today known as the Broad Street cholera outbreak) mark the birth of modern infectious disease epidemiology, and he is today considered one of the fathers of the discipline.

The hygienic conditions of London in the first half of the 19th century made the miasma theory of disease entirely plausible.

The sewage system was very poor, and the stench of animal and human feces, combined with that of rotting rubbish, was pervasive. According to old medical texts (which shared and propagated the beliefs and prejudices of the time), disease was more prevalent among poor people because they stank more and because their supposed moral corruption weakened their constitutions and made them more vulnerable to disease. The miasma theory, according to which cholera (along with several other diseases) was caused by "bad air," was dominant at the time, although a minority of scientists were skeptical. Two pamphlets had appeared within a few months of each other in 1849 that criticized the miasma theory: "On the Mode of Communication of Cholera," by John Snow, and "Malignant Cholera: its Mode of Propagation and its Prevention," by William Budd, physician to the Bristol Infirmary. Both essays contained a wealth of physiological and clinical observations supporting the theory of cholera as a waterborne contagious disease caused by a germ cell not yet identified. However, the epidemiological evidence required to substantiate this view was still lacking, and although William Budd claimed to have identified the agent of cholera in a fungus (a claim soon discredited), the miasma theory continued to be the subject of considerable debate and controversy during the years immediately after 1850.

In 1854, London was in the throes of an extremely vicious cholera epidemic, with about 500 people dying within the first 10 days. The epidemic broke out in the Soho district of London, close to John Snow's house. Snow mapped the 13 public wells and all the known cholera deaths around Soho, and he noted that the cases were spatially clustered around one particular water pump situated at the southwest corner of the intersection of Broad (now Broadwick) Street and Cambridge (now Lexington) Street (Fig. 1). London's water supply system consisted of shallow public wells where people could pump their own water to carry home, and about a dozen water utilities that drew water from the Thames to supply a jumble of water lines to more up-scale houses. Snow examined water samples from various wells with a microscope and confirmed the presence of "white, flocculent particles" in the Broad Street samples. Despite strong skepticism from the local authorities, he managed to have the pump handle removed from the Broad Street pump, and the spread of cholera declined soon.

Snow subsequently published a map of the epidemic to support his theory, showing the locations of the 13 public wells in the area and the 578 cholera deaths mapped by home address (Fig. 1). Snow used some proto-GIS (geographical information system) methods to support his argument; first, he drew Thiessen polygons around the wells, defining straight-line, least-distance service areas for each. A large majority of the cholera deaths fell within the Thiessen polygon surrounding the Broad Street pump, and a large portion of the remaining deaths were on the Broad Street side of the polygon surrounding the bad-tasting Carnaby Street well. In contrast, the large workhouse north of Broad Street experienced very few cholera deaths because it had its own well. Next, Snow redrew the service area polygons to reflect the shortest routes along streets to wells, and an even larger proportion of the cholera deaths fell within the shortest-travel-distance area around the Broad Street pump.

John Snow's investigation of the 1854 cholera epidemic in London was a brilliant lesson on how empirical knowledge could be efficiently translated into action for the well-being of everyone, but it was not enough to bring about a paradigm shift. Despite his attempts, Snow had failed to identify the germ causing cholera, and the scientific community continued to believe that cholera was an airborne disease.

THE DISCOVERY OF *VIBRIO CHOLERAE*

In the 19th century, there was much debate on cholera among European scientists. The

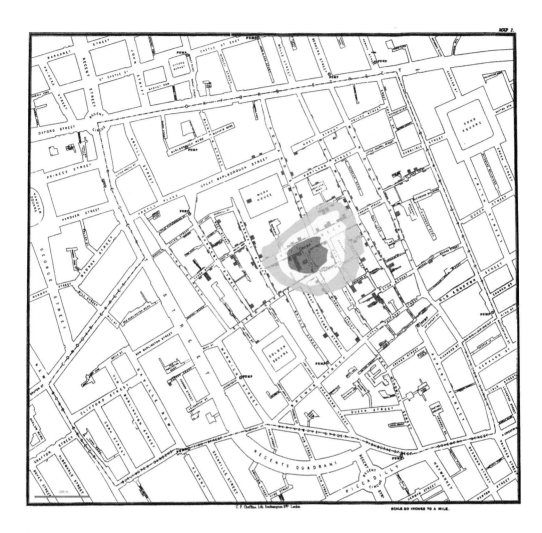

FIGURE 1 Map by John Snow of the cholera outbreak in Soho, London, 1854, modified by means of geoprofiling methods to show the areas most likely (from red to green) to have contained the source of infection. The pumps are marked with red circles, and the deaths from cholera are marked with blue squares. (Courtesy of Ugo Santosuosso and Alessio Papini, University of Florence, Florence, Italy.)

discussions centered on the merits of the two competing theories, the miasma theory and the germ theory. The usefulness of hygiene, sanitation, and quarantine in the control and prevention of the spread of cholera was debated as well. The most widely accepted explanation of the epidemics was based on the assumption that they were caused by a miasma, which was believed to be a harmful form of "bad air" or a poisonous vapor filled with particles from decomposed matter. This concept was later replaced by the scientifically founded germ theory of disease, based on the hypothesis that microorganisms can infect the body and cause specific diseases. Curiously, the germ responsible for cholera, *V. cholerae*, was discovered independently by two widely known and respected scientists. The first discovery took place in 1854, but the then-dominant miasma theory of diseases prevented its widespread acceptance by the scientific community. It was necessary to wait for almost 30 years until the infectious nature of the disease was finally established and accepted by everyone.

The First Discovery: Filippo Pacini, 1854

In 1854, the year of the notorious cholera outbreak in London investigated by John Snow, the disease also struck mainland Europe. Italy was not spared, and cholera outbreaks occurred in several cities and towns of the peninsula. Filippo Pacini (1812 to 1883) was a professor of anatomy and histology at the Istituto di Studi Superiori (Institute of Advanced Studies) of Florence in 1854, when the city fell into the grip of a terrible cholera outbreak. Pacini had developed a superlative mastery of the application of microscopy technique to anatomy, and he exploited his ability to identify for the first time the causative agent of cholera.

Pacini recorded a step-by-step description of the discovery of the vibrio in his journals, which are preserved in the Central National Library of Florence (8). During the cholera outbreak, he had the opportunity to examine the corpses of the patients who had died in the public hospital of Santa Maria Nuova in Florence and of the washerwomen who were in charge of cleaning the hospital linen. Pacini found in the stool samples and intestinal mucosa of people who had died of cholera millions of small bacilli with a comma shape, which he called vibrios. He described the disease as a massive loss of fluids and electrolytes and clearly stated that it was due to the local action of the vibrios in the intestinal mucosa, thus opposing the miasma theory in vogue at the time. In his own words, the vibrio "exists, can be seen, and it's not presumed," but is a real element of infection, "organic, living substance, of parasitic nature, communicating, reproducing, and then producing a disease of special character." He also recommended, in the most severe cases, the intravenous injection of sodium chloride diluted in water, a therapeutic measure that later proved very helpful in the treatment of cholera. Despite having identified the causative agent of the disease, Pacini's discovery of the vibrio was ignored by the scientific community. According to the miasma theory, influenced in turn by the localist/contagionist theory of the leading German scientist Max von Pettenkoffer, cholera was an airborne disease caused by a combination of three factors: bad or corrupted air (the so-called miasma), local and seasonal conditions, and a constitutional predisposition to infection.

In addition to his unrecognized contribution to the investigation of the causes of cholera, Pacini also developed a method to "resuscitate" people who were in a state of apparent death, not uncommon among those affected by cholera. Cholera victims were generally buried soon after death to prevent spread of the infection. Pacini noticed that when workers in the anatomical rooms lifted a corpse while holding it by the armpits and the feet, the cadaver would inhale deeply, then exhale while being placed on the anatomical table. Based on this observation, Pacini in 1870 developed a procedure to artificially stimulate respiration, consisting of a rhythmic mobilization of the upper limbs of an unconscious patient. He suggested placing the person in a supine position with the head lying on the raised end of an inclined surface and with the neck, chest, and abdomen uncovered. The mouth and throat were then examined in order to remove any obstacle to the passage of air through the airways, with care taken not to push anything into the stomach, which would trigger regurgitation and the passage of fluids into the airways. Finally, the operator had to lift the patient's shoulder girdle (i.e. the set of bones that connects the arm to the axial skeleton on each side) while pulling the patient toward himself. The movement had to pass through the clavicles to the sternum, causing the ribs to rise. Using this method, Pacini succeeded at reviving many cholera patients.

The Second Discovery: Robert Koch, 1883

In August 1883, Robert Koch traveled with a group of German colleagues from Berlin to Alexandria, Egypt, where cholera was epidemic. By that time, Koch was already a world-renowned scientist thanks to his identification of bacteria responsible for human diseases, such as *Bacillus anthracis* and *Mycobacterium tuberculosis*. Based on the work of the French scientist Louis Pasteur (1822 to 1895), Koch introduced new techniques and means to cultivate, manipulate, and characterize microorganisms. One of his most important innovations was to use solid media instead of liquid to prepare pure cultures of bacteria; liquid media are easily contaminated by other germs, and colonies of bacteria intermingle. With solid media, colonies can be kept isolated. Koch first grew his germs on sliced potatoes but later developed more sophisticated techniques, using agar gelatin in Petri dishes.

In Egypt, Koch and his colleagues Georg Gaffky and Bernhard Fischer carried out many postmortem examinations, finding a bacillus in the intestinal mucosa that was present only in the corpses of persons who had died of cholera. He deduced that the bacillus was related to the cholera process, but he was not sure whether it was causal or consequential.

Later in 1883, Koch sailed to Calcutta, India, where the epidemic was still very active. During a period of 24 weeks, Koch sent seven dispatches to the German Secretary of State for the Interior, which described the research in progress and were made available to the German press as they were received. On January 7th, 1884, Koch announced that he had successfully isolated the bacillus in pure culture and listed its characteristics and properties (8). However, he also had to admit that he had failed to reproduce the disease in animals, reasoning correctly that they are not susceptible but failing to present one of the elements of proof that he himself had

proposed. (These were later known as Koch's postulates, which provided a framework for proving the role of microbes in disease.)

Koch also understood the importance of clean water in preventing cholera, and the introduction of filtered water pipes led to a fall in the incidence of the disease. This added evidence in support of Koch's theory and showed a way to solve the problem by using the weapon of prevention. However, taking advantage of Koch's failure to reproduce cholera in animals, his rivals claimed that a causal relationship between the "comma" bacillus and cholera had not been proved. At the cholera conferences in Berlin (July 1884) and Rome (May 1885), opinions were not unanimously in favor of Koch's view of cholera as a transmissible disease. Times were changing rapidly, however, and the infectious theory of cholera soon prevailed, although it was necessary to wait until 1959 to obtain a definitive confirmation with the discovery of the toxin responsible for the disease.

AN EVER-PRESENT THREAT

Mortality from infectious diseases declined sharply in most areas of the world during the 20th century. Indeed, several infectious diseases are no longer a significant public health problem, either because they can be prevented with a vaccine or because they are easily treated with antibiotics. Smallpox has been eradicated, polio has been eliminated in many world regions, bubonic plague has nearly disappeared (although the reasons for its disappearance have not been completely elucidated), and syphilis is 10 times less common today than it was in the early 20th century and can be treated effectively in its early stages. Cholera, however, has not disappeared. The routes of transmission and predisposing factors are well-known, safe and effective vaccine exists, and infected people can be successfully treated with oral rehydration therapy. Nonetheless, a seventh cholera pandemic originated almost 50 years ago in Celebes; driven by the El Tor biotype, it has established endemicity in nearly every world region and has caused major epidemics periodically, such as those in Zimbabwe (2008), Haiti (2010), Sierra Leone (2012), Mexico (2013), and South Sudan and Ghana (2014) (9, 10).

The WHO figures on the morbidity and mortality of cholera are nothing less than impressive (11). In 2009, more than 220,000 cholera cases and approximately 5,000 cholera deaths (for a fatality rate exceeding 2%) were reported to the WHO from 45 countries worldwide. However, cholera is largely underreported (probably only 5% to 10% of actual cases are reported), especially in countries where it is most incident, and the WHO has estimated that 3 million to 5 million cases of cholera occur globally every year, with 100,000 to 120,000 deaths. Children younger than 5 years of age are the most vulnerable in countries where the disease is endemic, and nearly half of all cholera cases occur in this age group; in areas of cholera outbreaks and epidemics, however, cholera affects mostly adults, as has happened in the Americas since 1991. Even more worrying, the WHO estimated that the population at risk for cholera consists of a breathtaking 1.4 billion individuals; in other words, almost all developing countries face the threat of cholera epidemics.

The persistence of cholera as a public health problem worldwide certainly depends on the type of transmission (fecal–oral route, without carriers), but the main factors predisposing to its spread are socioeconomic rather than purely biological. Ecological studies are conclusive in showing that the incidence of cholera correlates positively with socioeconomic indicators such as a low gross national product per capita, a high proportion of householders with an income less than or equal to the minimum wage, and literacy rates below 90% (12). Briefly, cholera is an indicator of lagging social development, as the global burden of cholera (morbidity and mortality) is borne almost entirely by those countries where basic infrastructure is available for only a minority of the population. The areas that are typically most at risk for cholera outbreaks are high-density slums with poor living conditions and inadequate or no municipal services, often surrounding the inner cores of big cities (13). Cholera also flourishes in situations where a breakdown of already fragile sanitation and health infrastructure occurs because of natural disasters or humanitarian crises; among the several available examples, we are reminded of the cholera outbreak in a refugee camp in Kenya in 2009 and the epidemic that began in Haiti a few months after the earthquake of January 2010 (14).

For cholera epidemics to occur, a (quasi) sine qua non is the presence of unsatisfactory hygienic conditions, including a shortage of safe drinking water, insufficient sanitation, crowded housing, and the lack of an efficient sewage system. Briefly, cholera spreads and thrives where the overall social and environmental status is poor, whereas it no longer poses a threat (except for the possibility of sporadic episodes) to countries where minimum standards of hygiene are met. Like other diseases that mainly affect impoverished people, cholera is not only a public health problem; it has considerable economic costs as well and hinders a reduction in the health gap that separates developed and nondeveloped countries. Cholera can impact the economy of a country in many ways, including loss of the productivity of affected people and their caregivers, adverse effects on tourism, and embargoes on food trade with other countries.

Cholera was limited to a well-defined world region until the early 19th century. In an apparent paradox, it is more widespread today than it was two centuries ago despite the impressive advances in the fight against infectious diseases that were made during this period. Indeed, cholera was even able to reappear in areas where it had been eliminated many years before. Cholera has adapted very well to the social changes of the past two centuries, many of which favor its spread, such as the persistence of areas with poor hygienic conditions, frequent population displacements caused by war or natural disasters, and widening of the gap between rich and poor in many underdeveloped or developing countries. National governments and international organizations have had available for many years public health interventions that have proved to be effective against cholera; sadly, however, it is humankind, not cholera, that seems to fall back and lose its ground. Chroniclers and historians of cholera will still have to live side by side for several more years.

CITATION

Lippi D, Gotuzzo E, Caini S. 2016. Cholera. Microbiol Spectrum 4(4):PoH-0012-2015.

REFERENCES

1. **Heymann DL (ed).** 2008. *Control of Communicable Disease Manual*, 19th ed. American Public Health Association (APHA), Washington, DC.
2. **Lipp EK, Huq A, Colwell RR.** 2002. Effects of global climate on infectious disease: the cholera model. *Clin Microbiol Rev* **15:**757–770.
3. **Barua D.** 1992. History of cholera, p 1–35. *In* Barua D, Greenough WB (ed), *Cholera*. Plenum Publishing, New York, NY.
4. **Siddique AK, Cash R.** 2014. Cholera outbreaks in the classical biotype era. *Curr Top Microbiol Immunol* **379:**1–16.
5. **Lacey SW.** 1995. Cholera: calamitous past, ominous future. *Clin Infect Dis* **20:**1409–1419.
6. **Smith GD.** 2002. Commentary: Behind the Broad Street pump: aetiology, epidemiology and prevention of cholera in mid-19th century Britain. *Int J Epidemiol* **31:**920–932.
7. **Buechner JS, Constantine H, Gjelsvik A.** 2004. John Snow and the Broad Street pump: 150 years of epidemiology. *Med Health R I.* **87:**314–315.
8. **Lippi D, Gotuzzo E.** 2014. The greatest steps towards the discovery of *Vibrio cholerae. Clin Microbiol Infect* **20:**191–195.
9. **Poirier MJ, Izurieta R, Malavade SS, McDonald MD.** 2012. Re-emergence of cholera in the Americas: risks, susceptibility, and ecology. *J Glob Infect Dis* **4:**162–171.
10. **Morris JG Jr.** 2011. Cholera—modern pandemic disease of ancient lineage. *Emerg Infect Dis* **17:**2099–2104.
11. **Ali M, Lopez AL, You YA, Kim YE, Sah B, Maskery B, Clemens J.** 2012. The global burden of cholera. *Bull World Health Org* **90:** 209–218A.
12. **Talavera A, Pérez EM.** 2009. Is cholera disease associated with poverty? *J Infect Dev Ctries* **3:**408–411.
13. **Naseer M, Jamali T.** 2014. Epidemiology, determinants and dynamics of cholera in Pakistan: gaps and prospects for future research. *J Coll Physicians Surg Pak* **24:**855–860.
14. **Lantagne D, Balakrish Nair G, Lanata CF, Cravioto A.** 2014. The cholera outbreak in Haiti: where and how did it begin? *Curr Top Microbiol Immunol* **379:**145–164.

Human Lice in Paleoentomology and Paleomicrobiology

18

REZAK DRALI,[1] KOSTA Y. MUMCUOGLU,[2] and DIDIER RAOULT[1]

Lice (Insecta, Phthiraptera) are permanent obligate parasites of birds and mammals. Approximately 4,900 species of lice are recorded and distributed into four suborders: chewing or biting lice, including Rhynchophthirina, Ischnocera, and Amblycera, and sucking lice, Anoplura (1).

The recent discovery of two fossils, with estimated ages of 44 million years (2) and 100 million years (3), provides an indication about the origin of lice. These insects even survived the mass extinctions of species 65 million years ago, corresponding to the Cretaceous–Paleogene (K–Pg) boundary (4).

Among the 15 families of lice included in the suborder Anoplura, two are found in humans (5). The family Pediculidae contains the genus *Pediculus*, which is shared by humans and chimpanzees, and the family Pthiridae contains the genus *Pthirus*, which is shared by humans and gorillas (6). With the use of molecular clock analysis, a divergence time of 11.5 million years ago was found for the *Pthirus–Pediculus* split, and a divergence time of 5.6 million years ago was found for the split between *Pediculus schaeffi*, found in chimpanzees, and the human louse *Pediculus humanus* (7, 8). The age of the most recent common ancestor of the two *Pediculus* species studied matches the age

[1]Unité de Recherche sur les Maladies Infectieuses et Tropicales Emergentes: URMITE, Aix Marseille Université, UMR CNRS 7278, IRD 198, INSERM 1095, Faculté de Médecine, Marseille, France; [2]Parasitology Unit, Department of Microbiology and Molecular Genetics, The Kuvin Center for the Study of Infectious and Tropical Diseases, Hadassah Medical School, The Hebrew University, Jerusalem, Israel.

Paleomicrobiology of Humans
Edited by Michel Drancourt and Didier Raoult
© 2016 American Society for Microbiology, Washington, DC
doi:10.1128/microbiolspec.PoH-0005-2014

predicted by the host divergence (approximately 6 million years), whereas the age of the ancestor of *Pthirus* does not. The two species of *Pthirus* (*Pthirus gorillae* and *Pthirus pubis*) last shared an ancestor approximately 3 million to 4 million years ago, which is considerably later than the divergence between their hosts (gorillas and humans, respectively) at approximately 7 million years ago. This would be the result of host switching from archaic gorillas to archaic hominids at roughly 3 million years ago via direct contact between them (9).

Population expansion in human lice is coincident with the out-of-Africa expansion of humans (100,000 years ago) (54). Molecular clock analysis of head and body lice indicates that body lice originated 72, 000 ± 42,000 years ago, which correlates with the expansion of humans out of Africa and the wearing of clothing for protection from the colder climatic conditions (10).

MODERN HUMAN LICE

P. humanus humanus, the body louse, and *P. humanus capitis*, the head louse, are two louse ecotypes, each of which occupies an ecological niche in its host. The head louse lives and breeds in the hair of the head, whereas the body louse lives in clothing, where it lays its eggs in the seams and folds (11).

Pediculosis due to the body louse affects exclusively precarious populations, such as the homeless, prisoners, and war refugees (12). In contrast, head lice preferentially infest schoolchildren, with hundreds of millions of cases reported each year worldwide, regardless of hygienic conditions (13).

For some time, various comparative studies were unable to differentiate between these two types of lice. As recently as 1919, Nuttall designated body lice and head lice as two ecotypes representing extreme variations of the same species because they could not be distinguished in all essential points of structure (14). However, the advent of molecular biology allowed the use of several genetic markers in the analysis of lice in an attempt to find answers to the various remaining questions.

The analysis of mitochondrial genes allowed the classification of *P. humanus* into three different clades, designated clades A, B, and C. Only clade A comprises both head and body lice and is distributed worldwide (7). Clade B contains head lice found in the Americas, western Europe, Australia, and North Africa, whereas clade C contains head lice found in Nepal, Ethiopia, and Senegal (15). The study of lice recovered from pre-Columbian mummies made it possible to determine that clades A and B were present before the arrival of European settlers in the Americas (16, 17), reinforcing the assessment that lice belonging to clade B were of American origin and were subsequently introduced into the Old World through the return of European settlers (15). A fourth mitochondrial clade, clade D, is found in Africa, in the Democratic Republic of the Congo. Similar to clade A, clade D contains both body and head lice (Fig. 1).

Regarding nuclear genes, the phylogenetic analysis of 18S rRNA sequences showed that human lice can be classified into those from sub-Saharan Africa and those from other regions (18). Conversely, analyses targeting elongation factor-1α and RNA polymerase II genes showed that there is more diversity in African than in non-African lice, and more diversity in head lice than in body lice (10).

Two studies based on the use of intergenic spacers have revealed existing associations between the geographical sources and genotypic distributions of lice (19, 20). It has also been suggested that body louse populations may emerge from local head louse populations under poor hygienic conditions (20).

After two centuries of debate about the ecological and genotypic status of head and body lice, the comparison of their transcriptional profiles showed that both ecotypes have the same number of genes, with the

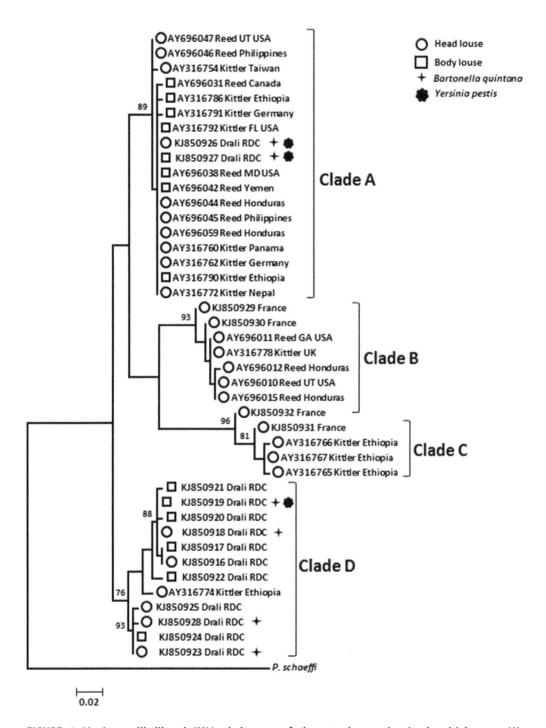

FIGURE 1 Maximum likelihood (ML) phylogram of the cytochrome *b* mitochondrial gene. ML bootstrapping supporting values greater than 75 are located above the nodes. Mitochondrial clade memberships are indicated to the right of each tree. GenBank accession numbers, manuscript lead author, and locality are indicated for each louse specimen. Localities are abbreviated as follows: Florida, FL; Georgia, GA; Maryland, MD; Democratic Republic of the Congo, RDC; United Kingdom, UK; Utah, UT.

exception of one gene that is missing in the head louse (PHUM540560) (21). Subsequently, analysis of a portion of this gene in 142 head and body lice belonging to mitochondrial clade A, collected from mono-infested hosts from 13 countries on five continents, showed that the PHUM540560 gene is present in both ecotypes. However, 22 polymorphisms were characterized between the sequences of body and head lice, allowing the development of a valuable tool using multiplex reverse transcription-PCR (RT-PCR) to differentiate quickly between the two ecotypes (22). Now, it is possible to differentiate between head and body lice, including those collected from the head and clothing of dually infested individuals (23).

To date, only the body louse is considered to be the vector of three dangerous diseases that have ravaged entire populations throughout history (24): epidemic typhus, trench fever, and relapsing fever, which are caused by *Rickettsia prowazekii*, *Bartonella quintana*, and *Borrelia recurrentis*, respectively (25). Body lice are also suspected in the transmission of *Yersinia pestis*, the agent of plague (26–28).

In recent years, the DNA of *B. quintana* was detected in head lice belonging to clade A (12, 28–30) and clade C (31, 32), whereas the DNA of *B. recurrentis* was found in head lice belonging to clade C (33) and the DNA of *Y. pestis* was detected in head lice of clade A and clade D (Fig. 1).

HUMAN LICE FROM ARCHAEOLOGICAL EXCAVATIONS AND HUMAN BODY REMAINS

The three human lice (i.e., head, body, and pubic lice), together with the two follicle mites, *Demodex folliculorum* and *Demodex brevis*, are most likely the oldest permanent ectoparasites of humans. Lice are mentioned in the Bible as the third plague visited on the Egyptians when Pharaoh denied the request of Moses to let the Israelites go (Exodus 8:16). In the 16th century BC, an Egyptian text known as the Papyrus Ebers described a remedy for lice prepared from date flour.

In recent decades, ancient lice and nits have been recovered in excavations and human body remains from different archaeological sites around the world (Fig. 2). The first ancient lice were isolated from Egyptian mummies in 1924 by Ewing (34). Later, head lice and eggs were found on mummies in China (35), Greenland (36–39), the Aleutian Islands (40), and South America (16, 17, 34, 41–46).

Head louse eggs were recovered on hair from human remains found in Brazil and were carbon-dated to approximately 10,000 years (41); in addition, 9,000-year-old louse eggs were found on hair samples from an individual who lived in Nahal Hemar Cave near the Dead Sea in Israel (47).

Head louse combs, very similar to modern louse combs, were already being used for delousing in Egypt in Pharaonic times (approximately 6,500 years ago) (48). Seven head lice were recovered from the debris found among the fine teeth of a wooden comb excavated in Antinoë, Egypt, and were dated to between the 5th and 6th centuries AD (49).

Head lice and their eggs were found in 12 of 24 hair combs recovered from archaeological excavations in the Judean and Negev Deserts of Israel, including from Masada and Qumran (50, 51). More recently, the head and apical part of one of the legs of a head louse were found in a wooden louse comb from the Roman period excavated in the "Cave of the Pool," close to the En Gedi Oasis near the Dead Sea (52). A head louse egg dating to the 1st century BC was also found in a louse comb excavated in the Christmas Cave in the Qidron Valley near the Dead Sea and Qumran (53).

Most of the combs found in archaeological excavations are made of wood, although some are made from bones and ivory. Most of the combs have two sides: one side to open hair knots and another to remove lice and

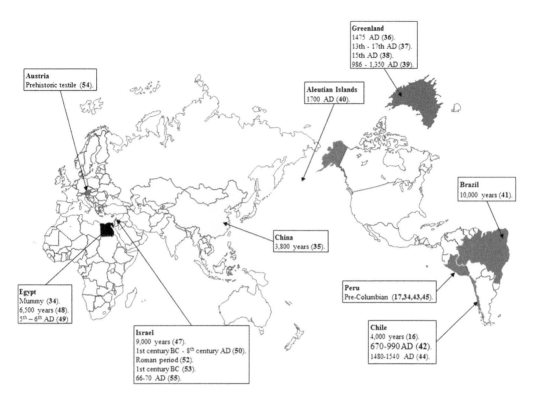

FIGURE 2 Map showing major areas of ancient lice recovery worldwide. The numbers in parentheses refer to references.

eggs. It can be assumed that these handmade combs were very effective in mechanically removing lice and eggs because they were not smooth or flexible; however, they were therefore more painful to use.

Body lice eggs were found in a prehistoric textile from Hallstaetter / Salzberg in Austria (54) and from deposits of farmers in Viking Greenland, dated to 986 to 1350 AD (39).

The remains of a body louse were also found in one of the rooms (the Casemate of the Scrolls) at the fortress of Masada. Originally constructed during the last decade of King Herod's reign, the casemate room was converted into a dwelling unit during the first Jewish revolt against the Romans. Following the conquest of Masada, the room was used by Roman soldiers as a dumping area. The context of the textiles associated with the louse and their nature clearly suggest an origin at the time of the rebellion (55).

Ancient specimens of pubic lice were found from a 2,000-year-old South American mummy (56), in human remains (mid-1st or 2nd century AD) from the Roman era in Britain (57), from post-medieval periods in Iceland, and in 18th century London (58–60).

MOLECULAR ANALYSIS OF ANCIENT LICE

Since the 1950s, paleoparasitological studies have contributed significantly to the understanding of parasite evolution and ecology (61). Because lice are obligatory parasites that complete all stages of their development in hosts, they are excellent markers of the evolution of humans (7). Ancient lice found in mummies and buried human remains are particularly useful in providing valuable information for tracking parasite and host migrations over time. Although studies based on molecular analyses of ancient lice are not numerous, it is clear that the results obtained are amazing.

In 2008, by analyzing two mitochondrial genes (*cytb* and *cox1*) in 1,000-year-old head lice collected from a Chiribaya mummy in Peru, Raoult et al. showed that the most prevalent and well-distributed mitochondrial clade of human lice – namely, clade A – had a pre-Columbian presence in the Americas (17). By analyzing a partial sequence of the mitochondrial cytochrome *b* gene in two operculated nits collected from a 4,000-year-old Chilean mummy, Boutellis et al. in 2013 showed that head lice belonging to clade A and clade B, which predated the arrival of European settlers in the Americas, can live in sympatry (16). In 2015, Drali et al.

developed a tool based on real-time PCR to test ancient head louse eggs recovered in Israel and dating from the Chalcolithic period (fourth millennium BC) and the early Islamic period (650 to 810 AD) (22). The phylogenetic analyses of these head louse egg sequences showed that they most likely were associated with people originating in West Africa because they belonged to the mitochondrial subclade specific to that region (Fig. 3).

MOLECULAR ANALYSIS OF ANCIENT LICE ASSOCIATED WITH BACTERIA

Proving the presence of louse-borne pathogens in ancient lice can help to determine the cause of death when a mass grave is discov-

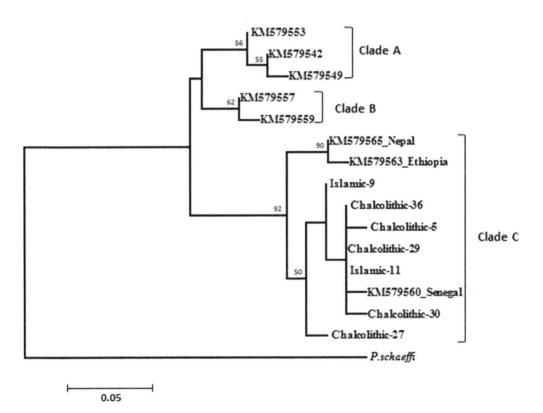

FIGURE 3 Maximum-likelihood (ML) phylogram of the cytochrome *b* mitochondrial gene. ML bootstrapping supporting values greater than 50 are located above the nodes. Mitochondrial clade memberships are indicated to the right of each tree.

ered. The most broadly used approach for achieving this is the detection and characterization of DNA (62). Host-associated microbial DNA can persist for 20,000 years (63), and bacterial DNA preserved in permafrost specimens has been dated to up to 1 million years (64).

Thus, in 2006, Raoult et al. concluded that the louse-borne infectious diseases that affected approximately one-third of Napoleon's soldiers buried in Vilnius, Lithuania, might have been a major cause of mortality during the French retreat from Russia (65). DNA from *B. quintana*, the agent of trench fever, was detected by PCR and sequenced in three of five body lice segments identified morphologically and molecularly by targeting a portion of the mitochondrial reduced nicotinamide adenine dinucleotide (NADH) dehydrogenase subunit 4 (*ND4*) gene. Similarly, DNA of *B. quintana* was found in the dental pulp of seven of 35 soldier remains, and the DNA of *R. prowazekii*, the agent of typhus, was found in another three soldiers (65).

In other studies, pathogens transmitted by lice have been identified successfully in the dental pulp of human remains from other sites and different periods. However, the presence of lice at archaeological sites from which human remains were extracted has not been reported. Because it is known that human infestation by lice was almost continuous during past periods (24), one might suspect that the absence of these parasites at archaeological sites is due to their degradation. For example, in 1998, Drancourt al. detected *Y. pestis* DNA in the dental pulp of a human who died 400 years ago, at a time when the plague was raging in France (66). In 2005, the DNA of *B. quintana* was found in the dental pulp of a person who died 4,000 years ago (67). DNA of *B. quintana* and *R. prowazekii* was also detected in ancient remnants of bodies from graves in Douai, France. This result supports that typhus and trench fever were involved in the decimation of the besiegers of Douai (1710 to 1712) during the War of the Spanish Succession (68).

CONCLUSION

The study of ancient parasites provides an excellent opportunity to access additional information in time and space, facilitating the understanding of parasite–host relationships. Identifying only the present configurations of genetic diversity or geographical distribution is not sufficient for understanding all the events that have marked this association. The main objective of paleoparasitology is to forge links between the past and present.

Thus, as we have observed through the study of ancient lice, it is possible to trace the migratory movement of *Homo sapiens* through the centuries, and it is also possible to learn more about the circulation of pathogens transmitted by lice. Lice have proved to be an exceptional record that allows the construction of credible scenarios, so that what occurred in the past can be understood upon the discovery of mass graves.

CITATION

Drali R, Mumcuoglu KY, Raoult D. 2016. Human lice in paleoentomology and paleomicrobiology. Microbiol Spectrum 4(3): PoH-0005-2014.

REFERENCES

1. **Johnson KP, Yoshizawa K, Smith VS.** 2004. Multiple origins of parasitism in lice. *Proc Biol Sci* **271:**1771–1776.
2. **Wappler T, Smith VS, Dalgleish RC.** 2004. Scratching an ancient itch: an Eocene bird louse fossil. *Proc Biol Sci* **271**(Suppl 5)**:**S255–S258.
3. **Grimaldi D, Engel MS.** 2006. Fossil Liposcelididae and the lice ages (Insecta: Psocodea). *Proc Biol Sci* **273:**625–633.
4. **Smith VS, Ford T, Johnson KP, Johnson PC, Yoshizawa K, Light JE.** 2011. Multiple lineages of lice pass through the K-Pg boundary. *Biol Lett* **7:**782–785.
5. **Light JE, Smith VS, Allen JM, Durden LA, Reed DL.** 2010. Evolutionary history of mammalian sucking lice (Phthiraptera: Anoplura). *BMC Evol Biol* **10:**292.
6. **Price MA, Graham OH.** 1997. *Chewing and Sucking Lice as Parasites of Mammals and*

Birds. Technical Bulletin No. 1849. United States Department of Agriculture, Beltsville, MD, USA.

7. **Reed DL, Smith VS, Hammond SL, Rogers AR, Clayton DH.** 2004. Genetic analysis of lice supports direct contact between modern and archaic humans. *PLoS Biol* **2:**e340.

8. **Stauffer RL, Walker A, Ryder OA, Lyons-Weiler M, Hedges SB.** 2001. Human and ape molecular clocks and constraints on paleontological hypotheses. *J Hered* **92:**469–474.

9. **Reed DL, Light JE, Allen JM, Kirchman JJ.** 2007. Pair of lice lost or parasites regained: the evolutionary history of anthropoid primate lice. *BMC Biol* **5:**7.

10. **Kittler R, Kayser M, Stoneking M.** 2003. Molecular evolution of *Pediculus humanus* and the origin of clothing. *Curr Biol* **13:**1414–1417.

11. **De Geer C.** 1778. *Mémoires pour servir à l'histoire des Insectes.* Hesselberg, Stockholm, Sweden.

12. **Sangare AK, Boutellis A, Drali R, Socolovschi C, Barker SC, Diatta G, Rogier C, Olive MM, Doumbo OK, Raoult R.** 2014. Detection of *Bartonella quintana* in African body and head lice. *Am J Trop Med Hyg* **91:**294–301.

13. **Chosidow O.** 2000. Scabies and pediculosis. *Lancet* **355:**819–826.

14. **Nuttall GH.** 1919. The systematic position, synonymy and iconography of *Pediculus humanus* and *Phthirus pubis. Parasitology* **11:**329–346.

15. **Boutellis A, Abi-Rached L, Raoult D.** 2014. The origin and distribution of human lice in the world. *Infect Genet Evol* **23:**209–217.

16. **Boutellis A, Drali R, Rivera MA, Mumcuoglu KY, Raoult D.** 2013. Evidence of sympatry of Clade A and Clade B head lice in a pre-Columbian Chilean mummy from Camarones. *PLoS One* **8:** e76818. doi:10.1371/journal.pone.0076818.

17. **Raoult D, Reed DL, Dittmar K, Kirchman JJ, Rolain JM, Guillen S, Light JE.** 2008. Molecular identification of lice from pre-Columbian mummies. *J Infect Dis* **197:**535–543.

18. **Yong Z, Fournier PE, Rydkina E, Raoult D.** 2003. The geographical segregation of human lice preceded that of *Pediculus humanus capitis* and *Pediculus humanus humanus. C R Biol* **326:** 565–574.

19. **Ascunce MS, Toups MA, Kassu G, Fane J, Scholl K, Reed DL.** 2013. Nuclear genetic diversity in human lice (*Pediculus humanus*) reveals continental differences and high inbreeding among worldwide populations. *PLoS One* **8:**e57619. doi:10.1371/journal.pone.0057619.

20. **Li W, Ortiz G, Fournier PE, Gimenez G, Reed DL, Pittendrigh B, Raoult D.** 2010. Genotyping of human lice suggests multiple emergencies of body lice from local head louse populations. *PLoS Negl Trop Dis* **4:**e641. doi:10.1371/journal.pntd.0000641.

21. **Olds BP, Coates BS, Steele LD, Sun W, Agunbiade TA, Yoon KS, Strycharz JP, Lee SH, Paige KN, Clark JM, Pittendrigh BR.** 2012. Comparison of the transcriptional profiles of head and body lice. *Insect Mol Biol* **21:**257–268.

22. **Drali R, Boutellis A, Raoult D, Rolain JM, Brouqui P.** 2013. Distinguishing body lice from head lice by multiplex real-time PCR analysis of the Phum_PHUM540560 gene. *PLoS One* **8:**e58088. doi:10.10.1371/journal. pone.0058088.

23. **Drali R, Sangare AK, Boutellis A, Angelakis E, Veracx A, Socolovschi C, Brouqui P, Raoult D.** 2014. *Bartonella quintana* in body lice from scalp hair of homeless persons, France. *Emerg Infect Dis* **20:**907–908.

24. **Zinsser H.** 1935. *Rats, lice and history.* Little Brown and Company, Boston, MA.

25. **Raoult D, Roux V.** 1999. The body louse as a vector of reemerging human diseases. *Clin Infect Dis* **29:**888–911.

26. **Blanc G, Baltazard M.** 1941. Recherches expérimentales sur la peste. L'infection du pou de l'homme, *Pediculus corporis* de Geer. *C R Acad Sci* **213:**849–851.

27. **Houhamdi L, Raoult D.** 2006. Experimental infection of human body lice with *Acinetobacter baumannii. Am J Trop Med Hyg* **74:**526–531.

28. **Piarroux R, Abedi AA, Shako JC, Kebela B, Karhemere S, Diatta G, Davoust B, Raoult D, Drancourt M.** 2013. Plague epidemics and lice, Democratic Republic of the Congo. *Emerg Infect Dis* **19:**505–506.

29. **Bonilla DL, Kabeya H, Henn J, Kramer VL, Kosoy MY.** 2009. *Bartonella quintana* in body lice and head lice from homeless persons, San Francisco, California, USA. *Emerg Infect Dis* **15:**912–915.

30. **Boutellis A, Veracx A, Angelakis E, Diatta G, Mediannikov O, Trape JF, Raoult D.** 2012. *Bartonella quintana* in head lice from Senegal. *Vector Borne Zoonotic Dis* **12:**564–567.

31. **Angelakis E, Diatta G, Abdissa A, Trape JF, Mediannikov O, Richet H, Raoult D.** 2011. Altitude-dependent *Bartonella quintana* genotype C in head lice, Ethiopia. *Emerg Infect Dis* **17:**2357–2359.

32. **Sasaki T, Poudel SKS, Isawa H, Hayashi T, Seki S, Tomita T, Sawabe K, Kobayashi M.** 2006. First molecular evidence of *Bartonella quintana* in *Pediculus humanus capitis* (Phthiraptera: Pediculidae) collected from Nepalese children. *J Med Entomol* **43:**110–112.

33. **Boutellis A, Mediannikov O, Bilcha KD, Ali J, Campelo D, Barker SC, Raoult D.** 2013. *Borrelia recurrentis* in head lice, Ethiopia. *Emerg Infect Dis* **19:**796–798.

34. **Ewing HE.** 1924. Lice from human mummies. *Science* **60:**389–390.

35. **Wen T, Zhaoyong X, Zhijie G, Yehua X, Jianghua S, Zhiyi G.** 1987. Observation on the ancient lice from Loulan. *Investigatio et Studium Naturae (Museum Historiae Naturae, Shanghaiense)* **7:**152–155.

36. **Bresciani J, Haarløv N, Nansen P, Moller G.** 1989. Head lice in mummified Greenlanders from AD 1475, p 89–92. *In* Hart Hansen JP, Gullov HC (ed), *The Mummies from Qilakitsoq—Eskimos in the 15th Century.* Meddelelser om Grønland, Man & Society, **vol 12.** Museum Tuscalanum Press, Copenhagen, Denmark.

37. **Forbes V, Dussault F, Bain A.** 2013. Contributions of ectoparasite studies in archaeology with two examples from the North Atlantic region. *Int J Paleopathol* **3:**158–164.

38. **Lorentzen B, Rørdam AM.** 1989. Investigation of faeces from a mummified Eskimo woman, p 139–143. *In* Hart Hansen JP, Gullov HC (ed), *The Mummies from Qilakitsoq—Eskimos in the 15th Century.* Meddelelser om Grønland, Man & Society, **vol 12.** Museum Tuscalanum Press, Copenhagen, Denmark.

39. **Sadler JP.** 1990. Records of ectoparasites on humans and sheep from Viking-age deposits in the former western settlement of Greenland. *J Med Entomol* **27:**628–631.

40. **Horne P.** 1979. Head lice from an Aleutian mummy. *Paleopathol News* **25:**7–8.

41. **Araujo A, Ferreira LF, Guidon N, Maues Da Serra FN, Reinhard KJ, Dittmar K.** 2000. Ten thousand years of head lice infection. *Parasitol Today* **16:**269.

42. **Arriaza B, Orellana NC, Barbosa HS, Menna-Barreto RF, Araujo A, Standen V.** 2012. Severe head lice infestation in an Andean mummy of Arica, Chile. *J Parasitol* **98:**433–436.

43. **Brothwell DR, Spearman R.** 1963. The hair of earlier peoples, p 426–436. *In* Brothwell DR, Higgs E (ed), *Science in Archaeology.* Thames and Hudson, London, UK.

44. **Horne PD, Kawasaki SQ.** 1984. The Prince of El Plomo: a paleopathological study. *Bull N Y Acad Med* **60:**925–931.

45. **Reinhard KJ, Buikstra J.** 2003. Louse infestation of the Chiribaya culture, southern Peru: variation in prevalence by age and sex. *Mem Inst Oswaldo Cruz* **98**(Suppl 1)**:**173–179.

46. **Rivera MA, Mumcuoglu KY, Matheny RT, Matheny DG.** 2008. Head lice eggs, *Anthropophthirus capitis,* from mummies of the Chinchorro tradition, Camarones 15-D, Northern Chile. *Chungara-Revista De Antropologia Chilena* **40:**31–39.

47. **Zias J, Mumcuoglu KY.** 1989. How the ancients de-loused themselves. *Bibl Archaeol Rev* **15:**66–69.

48. **Kamal H.** 1967. *A Dictionary of Pharaonic Medicine.* The National Publication House, Cairo, Egypt.

49. **Palma RL.** 1991. Ancient head lice on a wooden comb from Antinoe, Egypt. *J Egypt Archaeol* **77:**194.

50. **Mumcuoglu KY, Zias J.** 1988. Head lice, *Pediculus humanus capitis* (Anoplura: Pediculidae) from hair combs excavated in Israel and dated from the first century B.C. to the eighth century A.D. *J Med Entomol* **25:**545–547.

51. **Mumcuoglu KY, Zias J.** 1991. Pre-Pottery Neolithic B head lice found in Nahal Hemar Cave and dated 6,900–6,300 B.C.E. (uncalibrated). *Atikot* **20:**167–168.

52. **Mumcuoglu KY, Hadas G.** 2011. Head louse (*Pediculus humanus capitis*) remains in a louse comb from the Roman period excavated in the Dead Sea area of Israel. *Isr Expl J* **61:**223–229.

53. **Mumcuoglu KY, Gunneweg J.** 2012. A head louse egg, *Pediculus humanus capitis* found in a louse comb excavated in The Christmas Cave, which dates to the 1st c. B.C. and A.D. *In* Gunnweg J, Greenblatt C (ed), *Outdoor Qumran and the Dead Sea. Its Impact on the Indoor Bio- and Material Cultures at Qumran and the Judean Desert manuscript.* Proceedings of the joint Hebrew University and COST Action D-42 Cultural Heritage Workshop held at the Hebrew University of Jerusalem in May 25-26, 2010.

54. **Hundt HJ.** 1960. Vorgeschichtliche Gewebe aus dem Hallstaetter Salzberg. *Jahrbuch des Roemisch-Germanischen Zentralmuseums Mainz* **7:**126–141.

55. **Mumcuoglu KY, Zias J, Tarshis M, Lavi M, Stiebel GB.** 2003. Body louse remains in textiles excavated at Massada, Israel. *J Med Entomol* **40:**585–587.

56. **Rick FM, Rocha GC, Dittmar K, Coimbra CE Jr, Reinhard K, Bouchet F, Ferreira LF, Araujo A.** 2002. Crab louse infestation in pre-Columbian America. *J Parasitol* **88:**1266–1267.

57. **Buckland PC, Sadler JP, Sveinbjarnardóttir G.** 1992. Palaeoecological investigations at Reykholt, Western Iceland, p 149–168. *In* Morris C, Rackham D (ed), *Norse and Later Settlement and Subsistence in the North Atlantic.* Archetype Publications, Department of Archaeology, University of Glasgow, Glasgow, Scotland.

58. **Girling MA.** 1984. Eighteenth century records of human lice (Phthiraptera, Anoplura) and fleas (Siphonaptera, Pulicidae) in the City of London. *Entomol Mon Mag* **120:**207–210.

59. **Kenward H.** 1999. Pubic lice (*Pthirus pubis* L.) were present in Roman and medieval Britain. *Antiquity* **73:**911–915.

60. **Kenward H.** 2001. Pubic lice in Roman and medieval Britain. *Trends Parasitol* **17:** 167–168.

61. **Dittmar K.** 2009. Old parasites for a new world: the future of paleoparasitological research. A review. *J Parasitol* **95:**365–371.

62. **Drancourt M, Raoult D.** 2008. Molecular detection of past pathogens, p 55–68. *In* Raoult D, Drancourt M (ed), *Paleomicrobiology: Past Human Infections.* Springer-Verlag, Berlin, Germany.

63. **Willerslev E, Hansen AJ, Ronn R, Brand TB, Barnes I, Wiuf C, Gilichinsky D, Mitchell D, Cooper A.** 2004. Long-term persistence of bacterial DNA. *Curr Biol* **14:**R9–R10.

64. **Willerslev E, Hansen AJ, Poinar HN.** 2004. Isolation of nucleic acids and cultures from fossil ice and permafrost. *Trends Ecol Evol* **19:**141–147.

65. **Raoult D, Dutour O, Houhamdi L, Jankauskas R, Fournier PE, Ardagna Y, Drancourt M, Signoli M, La VD, Macia Y, Aboudharam G.** 2006. Evidence for louse-transmitted diseases in soldiers of Napoleon's Grand Army in Vilnius. *J Infect Dis* **193:**112–120.

66. **Drancourt M, Aboudharam G, Signoli M, Dutour O, Raoult D.** 1998. Detection of 400-year-old *Yersinia pestis* DNA in human dental pulp: an approach to the diagnosis of ancient septicemia. *Proc Natl Acad Sci U S A* **95:**12637–12640.

67. **Drancourt M, Tran-Hung L, Courtin J, Lumley H, Raoult D.** 2005. *Bartonella quintana* in a 4000-year-old human tooth. *J Infect Dis* **191:**607–611.

68. **Nguyen-Hieu T, Aboudharam G, Signoli M, Rigeade C, Drancourt M, Raoult D.** 2010. Evidence of a louse-borne outbreak involving typhus in Douai, 1710-1712 during the war of Spanish succession. *PLoS One* **5:**e15405. doi:10.1371/journal.pone.0015405.

Index